Harsh K. Gupta

Artificial Water Reservoir Triggered Seismicity

Focus on Koyna, India

Springer

Harsh K. Gupta
CSIR-National Geophysical Research Institute
Hyderabad, Telangana, India

ISBN 978-3-032-05872-0 ISBN 978-3-032-05873-7 (eBook)
https://doi.org/10.1007/978-3-032-05873-7

This Springer imprint is published by the registered company Springer Nature Switzerland AG
The registered company address is: Gewerbestrasse 11, 6330 Cham, Switzerland

To

Manju, my wife, daughters and their families

and

colleagues and friends all over the world who took interest in anthropogenic seismicity

and

Uma Anuradha for helping me for over two decades

Foreword

It is both an honour and a personal privilege to introduce this comprehensive reference volume *Artificial Water Reservoir Triggered Seismicity—Focus on Koyna, India,* authored by Professor Harsh K. Gupta.

Professor Gupta's name is inseparably linked to the study of Reservoir-Triggered Seismicity (RTS), a field he has shaped over decades through pioneering research, strategic leadership, and influential mentorship. From his early analyses at Koyna to his leadership roles—President of the International Union of Geodesy and Geophysics (IUGG), Vice President of the International Association of Seismology and Physics of the Earth's Interior (IASPEI) and Founding President of the Asian Seismological Commission—his impact has been profound.

Professor Gupta's career has evolved in parallel with the emergence of RTS as a critical topic in geoscience and hazard mitigation. His early investigations following the 1967 Koyna earthquake laid the groundwork for understanding how large reservoirs, under certain geological and hydrological conditions, can induce damaging earthquakes. These studies went beyond local hazard analysis, they provided criteria to discriminate RTS from natural seismicity, identified the characteristic features and underlying mechanisms, and inspired a new wave of research around the world. His long-standing efforts ultimately led to the launch of the International Continental Drilling Program (ICDP) at Koyna, enabling direct study of an earthquake source zone. His books, *Dams and Earthquakes* (1976) and *Reservoir Induced Earthquakes* (1992), became international benchmarks, while his *Encyclopedia of Solid Earth Geophysics* (2011, expanded in 2021) remains a landmark reference in the field.

As a respected leader within IUGG and IASPEI, and a dedicated advocate for scientific collaboration, Prof. Gupta played a key role in advancing the international recognition of triggered and induced seismicity as a distinct and pressing area of research. The proposal to establish the Triggered and Induced Seismicity Working Group (TAIS WG), was initiated and led by Prof. Stan Lasocki of the Institute of Geophysics, Polish Academy of Sciences. It gained momentum and visibility in large part, thanks to Prof. Gupta's strong endorsement and active support at the IASPEI level. His engagement was instrumental in securing approval of the initiative in 2007, which became a turning point for the community. The TAIS WG catalysed the organization of dedicated sessions at major assemblies and encouraged the emergence of new research infrastructures. It laid the foundation for a broader strategic vision that would ultimately lead to the creation of the Thematic Core Service Anthropogenic Hazards (TCS AH) and the EPISODES Platform. These collaborative environments linking data, tools, and expertise across institutional and national boundaries have enabled new forms of integration between researchers, engineers, and stakeholders. In many ways, Prof. Gupta's example of empowering initiatives, fostering interdisciplinary bridges, and mentoring the next generation has become a model for addressing the complex challenges of anthropogenic seismicity.

This book, coming nearly fifty years after the first systematic studies of RTS, offers a synthesis of knowledge that is both comprehensive and forward-looking. It begins with a general introduction and a global overview of 60 scientifically recognized RTS sites. Subsequent chapters explore the unique features of RTS sequences and the physical processes behind them, including pore pressure diffusion, elastic stress transfer, and fluid migration. A dedicated chapter focuses on high-magnitude RTS events with magnitudes over 6, such as those at Hsinfengkiang, Kariba,

Kremasta, and Polyphyto. At the heart of the book are two extensive chapters on Koyna, India—the most thoroughly studied RTS site, covering developments from 1962 to 2024, including borehole monitoring, short-term forecasting, and ICDP-supported scientific drilling. A comparative chapter on injection-induced seismicity in Oklahoma, USA, provides further insight into the broader spectrum of induced seismicity. The concluding chapter looks ahead, reflecting on how RTS sites like Koyna can address fundamental questions of earthquake nucleation, rupture, and hazard mitigation.

In bringing together these insights, the book advances scientific understanding and highlights the real-world relevance of RTS in a rapidly changing world. As nations continue to construct and manage large dams for water security, energy production, and development, RTS remains a multidisciplinary challenge involving science, engineering, and policy. The lessons distilled in these pages are vital for seismologists, engineers, planners, and all those entrusted with safeguarding communities.

My own first encounter with Prof. Gupta, though brief, left a lasting impression. I was a nervous young scientist presenting at the IUGG/IASPEI General Assembly in Cape Town, South Africa, in 2009. Pressed for time, two session chairs, both esteemed professors, Stan Lasocki and Andrzej Kijko, were about to cut my presentation short. To my immense relief, Prof. Gupta, seated beside me, intervened: "*Let her finish her presentation. She is our new generation.*" It was a simple gesture of fairness, but for me, it became a source of encouragement—a reminder that scientific progress depends on nurturing new voices and perspectives. I could not have imagined then that I would later have the privilege to collaborate closely with him, nor that his advocacy for the new generation would play out on a global scale.

I am confident that this volume will serve as both a valuable reference and a source of inspiration for researchers at every stage of their careers. Professor Gupta's legacy lies not only in his scientific discoveries, but in the spirit of rigorous science, open collaboration, and inclusive mentorship that he has fostered. As the field moves forward, facing new challenges and uncertainties, we move forward on the firm foundation laid by a true pioneer.

July 2025

Beata Orlecka-Sikora
Institute of Geophysics
Polish Academy of Sciences
Krakow, Poland

Preface

The objective of the present book is to provide an up-to-date knowledge about the artificial water reservoir triggered seismicity (RTS), globally, and inform about unique and outstanding research work having been carried out at the RTS site Koyna, India. The first book, entitled *Dams and Earthquakes* was published by Elsevier in 1976, authored by Harsh K. Gupta and B. K. Rastogi, was very successful. It was translated into the Russian (1978) and Chinese (1979) languages. An updated version of *Dams and Earthquakes* titled *Reservoir Induced Earthquakes*, authored by Harsh K. Gupta, was published in 1992 by Elsevier.

Artificial water reservoirs are constructed all over the world for flood control, power generation and irrigation. Under certain geological conditions, filling of these reservoirs has triggered earthquakes exceeding magnitude six, causing loss of human lives and damage to properties. The work carried over the last few decades has provided useful information as to why such triggered earthquakes occur and how safer dam sites could be selected. This book provides an up-to-date knowledge on all relevant issues for students, researchers, engineers and town planners.

Chapter 1 gives general introduction to RTS. Chapter 2 provides information about 53 sites globally where RTS events exceeding magnitude four have occurred. Chapter 3 deals with the characteristics of RTS sequences which discriminate them from the normal earthquake sequences. Chapter 4 deals with sites where RTS events of magnitude six and above have occurred. These include Hsinfengkiang, People's Republic of China; Kariba, Zambia-Zimbabwe border; Koyna, India; Kremasta, Greece; and Polyphyto, Greece. Additionally, the case of Lake Mead, USA, where the first artificial water reservoir triggered earthquake of magnitude five in 1939 was scientifically accepted. Whether the Sichuan earthquake of M 7.9 in 2008 in People's Republic of China was triggered by filling of the nearby Zipingpu Reservoir is debated. It is inferred not to be a case of RTS. Chapter 5 discusses the latest in understanding on the mechanism of RTS, including insights into the dynamics of RTS preparation process.

The RTS at Koyna, India is unique. Earthquakes started soon after impoundment of the Shivaji Sagar Lake in 1962. The most damaging triggered earthquake of M 6.3 occurred in Koyna, India in 1967. RTS at Koyna started soon after the impoundment of the Shivaji Sagar Reservoir in 1962. In 2005, the CSIR-NGRI, India set up a network of three matched components seismographs at seven locations and very detailed studies have been carried out on several aspects of RTS, which are unique globally.

Over the years the locale has hosted ~ 2 dozen M 5 and larger earthquakes and over 200 M 4 and larger earthquakes. Chapters 6 and 7 provide the total story of RTS at Koyna, including successful short-term M ~ 4 earthquake forecasts and setting up of a network of nine bore hole stations at a depth of ~ 1500 m. Globally, physical properties of rocks and fluids in the fault zone before, during and after the occurrence of an earthquake are lacking. Koyna has been found to be a suitable locale for such investigations as the earthquakes are confined to a region of 20 km X 30 km, earthquakes occur at a shallow depth (mostly between 2 and 9 km), the crust is cool (~ 130 degree Celsius at 6 km depth), there is no other source of earthquakes with in 50 km of Koyna Dam, the region is totally accessible for all kind of measurements. The suitability of Koyna for setting up of a deep borehole observatory was discussed during a ICDP workshop held in 2011 where some 60 experts participated and found the site to be suitable. Certain addition work such as LiDAR surveys, airborne gravity, and magnetic surveys, etc. were suggested. These were carried out during 2011–2014 and the second ICDP workshop held in 2014 discussed all the results and gave a go-ahead for setting up of a Pilot Borehole laboratory. The 3 km deep Pilot Borehole was completed in 2017, and all the results were discussed in 2017 ICDP workshop. The work is now continuing for putting up a 6–7 km borehole laboratory. For convenience the work carried out in Koyna is presented in two chapters: Chap. 6 dealing investigations from 1962 to 2004, and Chap. 7 from 2004 till end 2024. Chapter 8 deals with induced/triggered earthquakes in Oklahoma, USA, which is known to have low seismicity, based on historical records as well recent instrumental records. For enhancing oil production fluids were injected in deep borewells. An increase in seismicity was observed after about 20 years of initial fluid injection and about 5 years of the substantial increase in wellhead pressure including occurrence of M 5.7 and M 5.8 earthquakes in 2011 and 2016. Subsequently, fluid injection was curtailed. An investigation of the characteristics of the fluid injection induced earthquakes revealed that fluid injection induced earthquake sequences have characteristics like RTS. However, with so many wells operating simultaneously, the observed seismicity appears like earthquake swarms. The chapter deals with the fluid injection induced/triggered earthquakes in Oklahoma, USA.

Chapter 9, Way Forward, deals with what could be achieved in the years to come on RTS related issues. Globally, no earthquake model has been ever tested in the field. One of the problems is that in almost all earthquake bearing fault zones, mostly earthquakes occur at depths of 10 km or more and it has so far not been possible to carry out near-field studies of earthquakes to find the cause and effect of any physical parameter and nucleation of an earthquake. Hopefully, the study of triggered earthquakes at Koyna, India and elsewhere shall provide this much sought after information.

Interactions nationally and globally with many individuals have been very helpful. It is difficult to name all of them. Useful interactions are acknowledged with Uli Harms, Hans Kumpel, Brian Horsfield, and Johannes Schweitzer from Germany; V. P. Dimri, Mrinal Sen, V. M. Tiwari, R. K. Chadha, Prantik Mandal, K. Arora, B. P. K. Patro, N. P. Rao, Sukanta Roy, D. Shashidhar, H. V. S. Satyanarayana from CSIR-NGRI, Hyderabad, India; S. Nayak, M. Ravichandran, B. K. Bansal, S. Rajan and R. Mohan from the MoES, India; Stainslaw Lasocki and Beata Orlecka-Sikora from Poland; Pradeep Talwani, Art McGarr, David Simpson, and William Ellsworth from USA.

The successive Directors of the CSIR-NGRI, Hyderabad, Drs. V. P. Dimri, M. Sen, V. M. Tiwari and Prakash Kumar supported RTS related work, which is gratefully acknowledged. A special thanks to Uma Anuradha, Rajesh Rekapalli and K. Mallika for their help in the preparation of the manuscript of the book

I am grateful to Beata Orlecka-Sikora for her excellent Foreword.

I enjoyed working with Springer on this book.

Harsh K. Gupta
CSIR-NGRI
Hyderabad, India

Competing Interests The author has no competing interests to declare that are relevant to the content of this manuscript.

Contents

1 Introduction

Chapter 2 deals with the scientifically accepted cases of RTS globally where RTS events of magnitude 4 and above have occurred. This chapter also includes the physical factors that influence RTS. Beginning with the RTS observed at Lake Mead, USA in 1939, by the end of 2024, there are around 60 sites globally where RTS events exceeding magnitude 4 have occurred. At least at 5 sites RTS events have exceeded magnitude 6 and there are over a dozen sites where magnitude 5 has been exceeded. There have been publications where very well-established normal earthquakes have been included as the RTS events. This has created confusion. For a long time, a debate has continued as to what are the physical factors of the artificially filled water reservoirs that lead to triggering of earthquakes. Physical factors influencing the occurrence of RTS is also discussed in this chapter. Way back in 1976, it was observed that the height of the water column in the reservoir is the most important factor for triggering earthquakes. It was found that almost 26% of artificial water reservoirs which had water column height from 150 to 250 m triggered earthquakes. When the water column height is 120–150 m 17% of the reservoirs experienced triggered earthquakes. The percentage of reservoirs witnessing triggered earthquakes drops drastically to 6% when the water height is 90–120 m. This observation continues to hold good.

Chapter 3 discusses the common characteristics of RTS sequences. Around the beginning of 1970s there were around 20 known cases where, following the impoundment of an artificial water reservoir, a change in seismicity in the vicinity of the reservoir was observed. In most cases there was an increase in number and magnitude of the earthquakes occurring near the site of the water reservoir compared to the pre-impoundment stage. Very detailed investigations of the characteristics of the earthquake sequences occurring in the vicinity of the reservoir indicated that: (1) the b value in the earthquake magnitude–frequency relation ($\log N = A - b M$; where N is the number of earthquakes with magnitude M or larger, A and b are constants) are higher for the aftershocks and foreshocks as compared to the regional b values. (2) For RTS main events the decay of the aftershocks is slower compared to the normal earthquake sequences. (3) The magnitude ratio of the largest aftershock to main shock is higher. (4) The foreshock-aftershock pattern of the RTS events belongs to Type 2 of Mogi's earthquake models, whereas the regional normal earthquake sequences belong to Type 1 of the Mogi's model. It was realized that these characteristics relate to the mechanical properties of the media hosting the earthquakes. A deviation in these observed properties of the earthquake sequences indicates that impoundment of the reservoir has introduced heterogeneity in the Earth media hosting the earthquakes. In the years to follow all RTS earthquake sequences have exhibited similar characteristics.

In Chap. 4, the five sites, namely Hsinfengkiang, China; Kariba on Zambia-Zimbabwe border; Kremasta, Greece; Koyna, India; and Polyphyto, Greece, where triggered earthquakes of magnitude 6 or more have occurred are discussed. This includes the background of regional seismicity, general geology of the region, water impoundment history, and the observation and continuation of RTS. The characteristics of RTS sequences and how they differ from the normal earthquakes occurring in the nearby region is also investigated. Additionally, the case of Lake Mead, USA, where the first scientifically accepted case of artificial water reservoir triggered was identified way back in 1939, is reviewed. The case of M 7.9 Sichuan, China earthquake of 2008, whether it was triggered due to filling of the nearby Zipingpu Reservoir, is presented. Based on the characteristics of this earthquake sequence, it is not found to be an RTS event. The chapter would also include geological information of the Sichuan region.

Chapter 5 deals with the current knowledge of RTS mechanism. This chapter includes incremental stress due to water load, pore fluid pressure changes and its role in triggering earthquakes and all the other related issues. The work carried out at the Rocky Mountain Arsenal, Denver, Colorado, USA by Evans way back in 1966 laid the foundation to comprehend occurrence of RTS. He made use of Hubbert and Rubey's

H. K. Gupta, *Artificial Water Reservoir Triggered Seismicity*, https://doi.org/10.1007/978-3-032-05873-7_1

approach (Hubbert and Rubey 1959) to explain the role of increased pore fluid pressure in triggering earthquakes consequent to fluid injections. The three most important consequences of filling of an artificial water reservoir are: (1) Elastic stress increases following the filling of the reservoir; (2) There is an increase in pore fluid pressure; and (3) Pore pressure changes due to fluid migration. Recent work indicates that both pore pressure and Coulomb stress on the seismogenic fault contributed to the triggered earthquakes at Lake Aswan, Egypt. The chapter shall also deal with 'Induced or triggered earthquakes', and 'Rapid, Delayed and Continued seismicity'.

The most damaging triggered earthquake of M 6.3 occurred in Koyna, India in 1967. RTS at Koyna started soon after the impoundment of the Shivaji Sagar Reservoir in 1962. Over the years the locale has hosted ~ 2 dozen M 5 and larger earthquakes and over 200 M 4 and larger earthquakes. An immense amount of field surveys and research work has been carried out at Koyna. For convenience, RTS related work at Koyna has been dealt in two chapters. Chapter 6 deals with the period 1962 through 2004 and Chap. 7 for the period 2004 through 2024. These two chapters provide the total story of RTS at Koyna.

In Chap. 6, the initiation of RTS at Koyna after the impoundment of the Shivaji Sagar Lake is covered. Prior to 1962, there were no earthquakes reported from the Koyna region. A Milne Shaw seismograph had been operating at Pune since 1933, located about 110 km from Koyna Dam. It would have recorded any magnitude 3 and larger earthquake from the Koyna region. Earthquakes became prevalent after filling of the Shivaji Sagar lake in 1962. The M 6.3 earthquake on 10 December 1967 claimed around 200 human lives and destroyed the Koyna Nagar township. The source parameters of this earthquake were estimated. It was found to be a multiple seismic event. Interesting rotational movements caused by the earthquake were documented. The geological and geophysical investigation carried in the region are briefly presented. A detailed study of the water levels variation and its influence on RTS is presented. Among other findings, it was discovered that a weekly loading exceeding 40 ft/week was a necessary but not a sufficient condition for an M $\geq$ 5 earthquake to occur in the Koyna region. In 1985, Warna Reservoir located about 30 km south of Koyna was impounded. This provided a further impetus to RTS. An M 5.3 earthquake occurred in the immediate vicinity of the Warna Reservoir on 28 August 1993. A detailed study of earthquake focal mechanism revealed that the earthquakes occurring north of 17° 15′ latitude are basically strike-slip fault dominated, while on the south of 17° 15′, normal fault focal mechanism is prevalent.

Chapter 7 deals with the work having been carried out in the Koyna region during the period 2004–2024. The National Geophysical Research Institute, Hyderabad, India set up a network of 7 three component seismic stations in the vicinity of RTS area of Koyna and the seismic data were communicated to Hyderabad in near real time. M ~ 4 earthquakes were found to be preceded by swarms precursory to the main shock. Recognition of swarm preceding the main shock in real time led to making successful earthquake forecasts of M ~ 4 earthquakes. Globally, knowledge about the physical properties of rocks and fluids in the fault zone before, during and after the occurrence of an earthquake are lacking. Koyna has been found to be a suitable locale for such investigations as the earthquakes are confined to a region of 20 km × 30 km, earthquakes occur at a shallow depth (mostly between 2 and 9 km), the crust is cool (~ 130 °C at 6 km depth), there is no other source of earthquakes within 50 km of Koyna Dam, the region is totally accessible for all kind of measurements. With this in view, a detailed proposal was made to the International Continental Drilling Program (ICDP) in 2010. It was accepted and a dedicated workshop at NGRI, Hyderabad and field visit to Koyna were held in 2011. Attended by 74 experts, the workshop endorsed the proposal and suggested improving earthquake locations, carrying out airborne gravity and magnetic surveys, LiDAR and magnetotelluric surveys. All the suggested work was carried out over the next three years and reviewed in the next workshop held at Koyna in 2014. Satisfied with the progress of the work, the 2nd ICDP workshop gave a go ahead for putting a 3 km deep Pilot Borehole at an appropriate location. Based on earthquake distribution, logistic requirements and vicinity to the Donachiwada Fault that had hosted the 10 December 1967 M 6.3 and several M ~ 5 earthquakes, a location was chosen for the Pilot Borehole. The borehole was spudded in December 2016 and completed in June 2017. Several measurements were made during drilling and core samples from critical zones were collected for further studies. A Post-Operational ICDP workshop was held at Pune, India during 14–16 October 2017. It was attended by 67 participants including 19 from Canada, France, Germany, Italy, Japan, the Czech Republic, South Korea, Spain, Switzerland, and the USA. There was a total satisfaction with the results obtained and plans of further work leading to designing the proposed 7 km deep Main Borehole were discussed. Participants of the workshop visited the Borehole Geophysics Research Laboratory at Karad. The Core Repository was of immense interest. On the final day, detailed recommendations for the future course of action were made focusing on developing of Koyna RTS information system; characterization of Donachiwada fault; deployment of a vertical string of seismometers below 1500 m in the Pilot Borehole; a variety of hydrogeological/geophysical/geological measurements to be made; detailed study of the core samples; setting up of a reference borehole outside Koyna RTS area; and preparation of a detailed plan of the proposed deep bore laboratory at a depth of ~ 7 km. Overall, the placing of the borehole at a very appropriate location and carrying out all the necessary measurements during the drilling of 3-km-deep borehole were appreciated. In the year 2017, the completion of 5 decades of RTS at Koyna, India, was marked with commemoration of the 10 December 1967 earthquake in the session on Anthropogenic Seismicity at Kobe, Japan during the joint assembly of the International Association of Seismology and Physics of the Earth's Interior (IASPEI) and the International Association of Geodesy (IAG) during 30 July–4 August 2017. The Anthropogenic Seismicity was the second most attended session at Kobe with over a dozen presentations dealing with RTS at Koyna, India. The Journal of Geological Society of India dedicated December 2017 issue with 25 articles dealing with different aspects of RTS at Koyna, India. A proposal for a session was made to the 2017 annual fall meeting of the American Geophysical Union, at New Orleans, Louisiana, entitled "Fifty Years since the World's Largest Reservoir Triggered Seismic Event, the 10 December 1967 M 6.3 Koyna Earthquake—Lessons Learnt". It was merged with the session on "Induced Seismicity in the United States and Canada". The proceedings of this session were published as a Special Issue of the Bulletin Seismological Society of America in November 2018 with the heading, "Fifty Years after the 1967 Koyna Earthquake—Lessons Learned about Reservoir Triggered Seismicity (RTS)" (Rao et al. 2018).

The work was going on in full swing after the third ICDP workshop when the COVID-19 pandemic put brakes on the progress and broke the momentum. The 3 years of uncertainty (2019–2022) has resulted in certain setbacks in meeting the planned targets. Some of the planned targets and the timelines were worked out keeping in view the conditions prevailing in the country and in the world. I am hopeful, that a dream of drilling a 7 km deep borehole right into the heart of the seismic zone to catch a *Live* earthquake may become a reality someday.

Chapter 8 deals with the injection induced seismicity observed at Oklahoma, USA. The area is known to have low seismicity, based on historical records as well as recent instrumental records. For enhancing oil production, fluids were injected in deep borewells. An increase in seismicity was observed after about 20 years of initial fluid injection and about 5 years of the substantial increase in well-head pressure including occurrence of M 5.7 and M 5.8 earthquakes in 2011 and 2016. Subsequently, fluid injection was curtailed. An investigation of the characteristics of the fluid injection induced earthquakes revealed that fluid injection induced earthquake sequences have characteristics like RTS. However, with so many wells operating simultaneously, the observed seismicity appears like earthquake swarms. The chapter deals with the fluid injection induced/triggered earthquakes in Oklahoma, USA.

Chapter 9 deals with what future holds and what could be achieved through studies of RTS. As mentioned earlier, globally, no earthquake model has been ever tested in the field. The problems are that in almost all earthquake bearing fault zones, earthquakes occur at depths of 10 km or more, there are a complex mixture of faults and sources contributing to accumulation of strains and ultimate occurrence of earthquakes. It is very difficult to decipher the cause and effect ultimately leading to comprehending the initiation and progress of faulting. Moreover, most earthquake belts are in tectonically active terrains and observation and instrumentation surrounding the earthquake epicenter becomes difficult. In the case of huge water reservoirs, when they are constructed in stable continental regions and in case, they witness RTS, it becomes much easier to monitor all aspects of earthquakes as it has been done at Koyna, India. Hopefully, the study of triggered earthquakes at Koyna, India and elsewhere shall ultimately provide information on the following important issues:

1. Fluid pressure and permeability within and adjacent to the fault zone.
2. Composition and origin of fault-zone fluid and gasses.
3. Variation of the stress orientation and magnitude across the fault zone.
4. Nucleation of earthquakes.
5. Propagation of earthquake rupture.
6. The role of water reservoirs in triggering the earthquakes.

It would be a great contribution to understand the genesis of earthquakes through the study of RTS, globally.

References

Hubbert MK, Rubey WW (1959) Role of fluid pressure in mechanics of overthrust faulting, 1. Bull Geol Soc Am 70:115–166

Rao NP, Arora K, Ponomarev A, Nascimento AF (2018) Fifty years after the 1967 Koyna earthquake—lessons learned about reservoir triggered seismicity (RTS). Introduction to the special issue. Bull Seismol Soc Am 108(5B):2903–2906

2 Worldwide Distribution of Reservoir Triggered Seismicity (RTS)

Gupta (2011, 2021) updated cases of RTS that were first documented in 1992. As of the end of 2024, following is the breakup of RTS experienced globally, classified in accordance with the largest magnitude event of the RTS sequence:

1. 5 RTS sites where the largest earthquake was of $M \geq 6$,
2. 12 RTS sites where the largest magnitude was 5.0–5.9,
3. 36 RTS sites where the largest magnitude was 4.0–4.9.

2.1 Important Factors that Influence Occurrence of RTS

From the early 1970s, there has been a global effort to find the correlates that influence RTS. Earlier Simpson (1976), Stuart-Alexander and Mark (1976), Baecher and Keeney (1982) tried to correlate the largest magnitude event in the RTS sequence with the duration of loading, volume of the water stored as well as with the maximum depth of the water column in the reservoir. Figures 2.1, 2.2 and 2.3 updated from Gupta (1992) depict these correlations. No significant correlation could be established between the magnitude of the largest RTS event at a reservoir site and the time interval between the initial loading and the occurrence of the main event as well as reservoir volume with the magnitude of the largest event (Figs. 2.1, 2.2, and 2.3).

During nineteen seventies, Stuart-Alexander and Mark (1976) discovered a close correlation in the height of the water column in the reservoir and the occurrence of RTS. They pointed out that that almost a quarter of the reservoirs, where the height of water column was 150 m and more, witnessed RTS. This dropped to 17% when the water column was 120–150 m, and 6% when the water column was 90–120 m (Fig. 2.4). In the subsequent years also it has been observed that typically when the water column height in the reservoir is 100 m or more, RTS occurred. It can be noted in Table 2.1 that all the sites where RTS of magnitude 5 and larger occurred had a water column height of 100 m or more, with the exception for Marathon Dam in Greece, and Warna Dam in India. In a recent review of RTS in Brazil, Sayão et al. (2020), 30 cases are catalogued. The authors observe that the occurrence of RTS increases significantly with the increase of the height of the water column in the reservoir. When the height of water is 50 m or less, only 2% reservoirs experience RTS. However, this percentage increases to 54% when the water column height exceeds 100 m (Table 2.2; Fig. 2.5).

H. K. Gupta, *Artificial Water Reservoir Triggered Seismicity*, https://doi.org/10.1007/978-3-032-05873-7_2

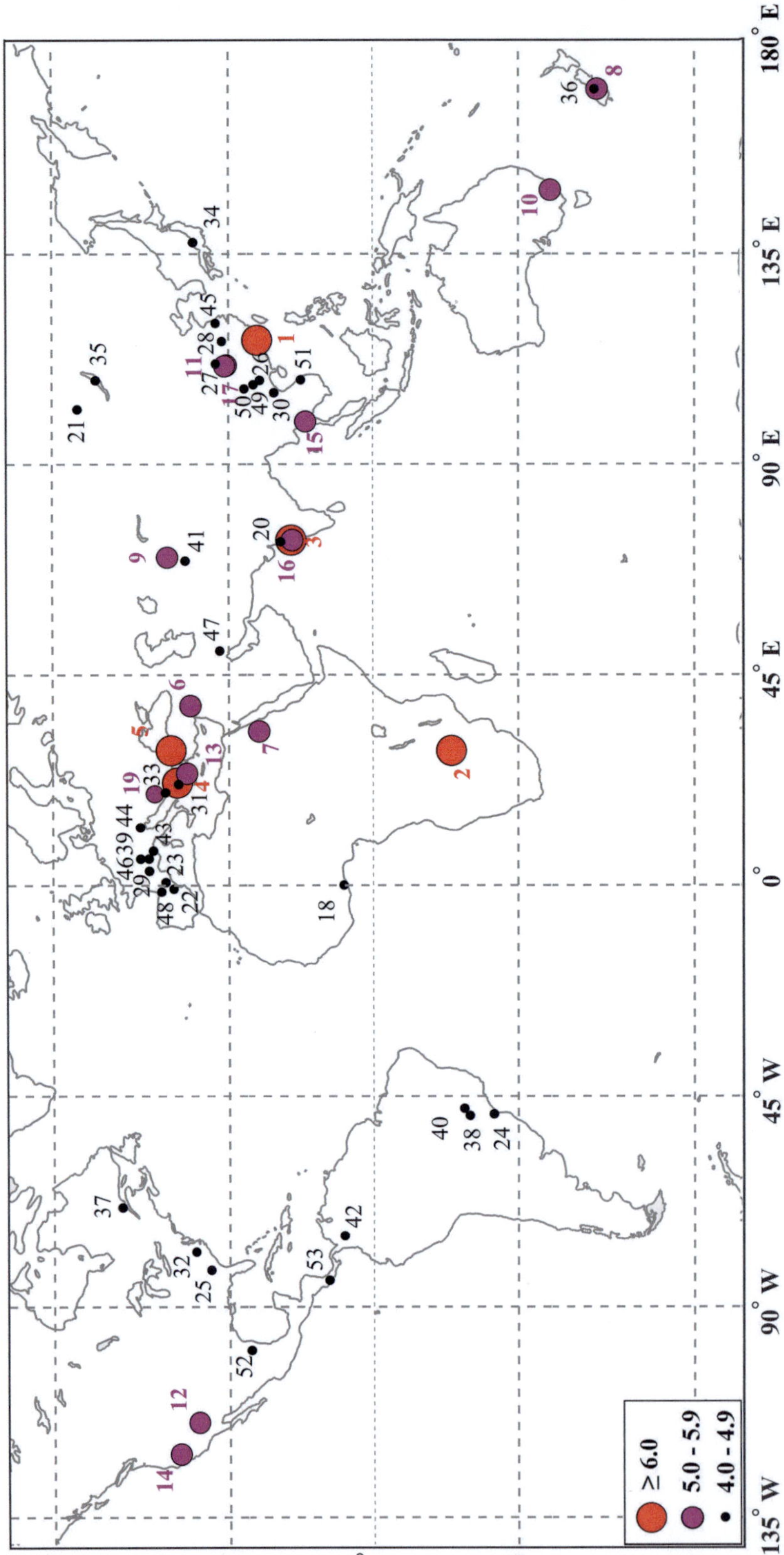

Fig. 2.1 Global distribution of RTS sites where the largest RTS event has been of magnitude 4 and above. Updated from Gupta (1992). Details are provided in Table 2.1

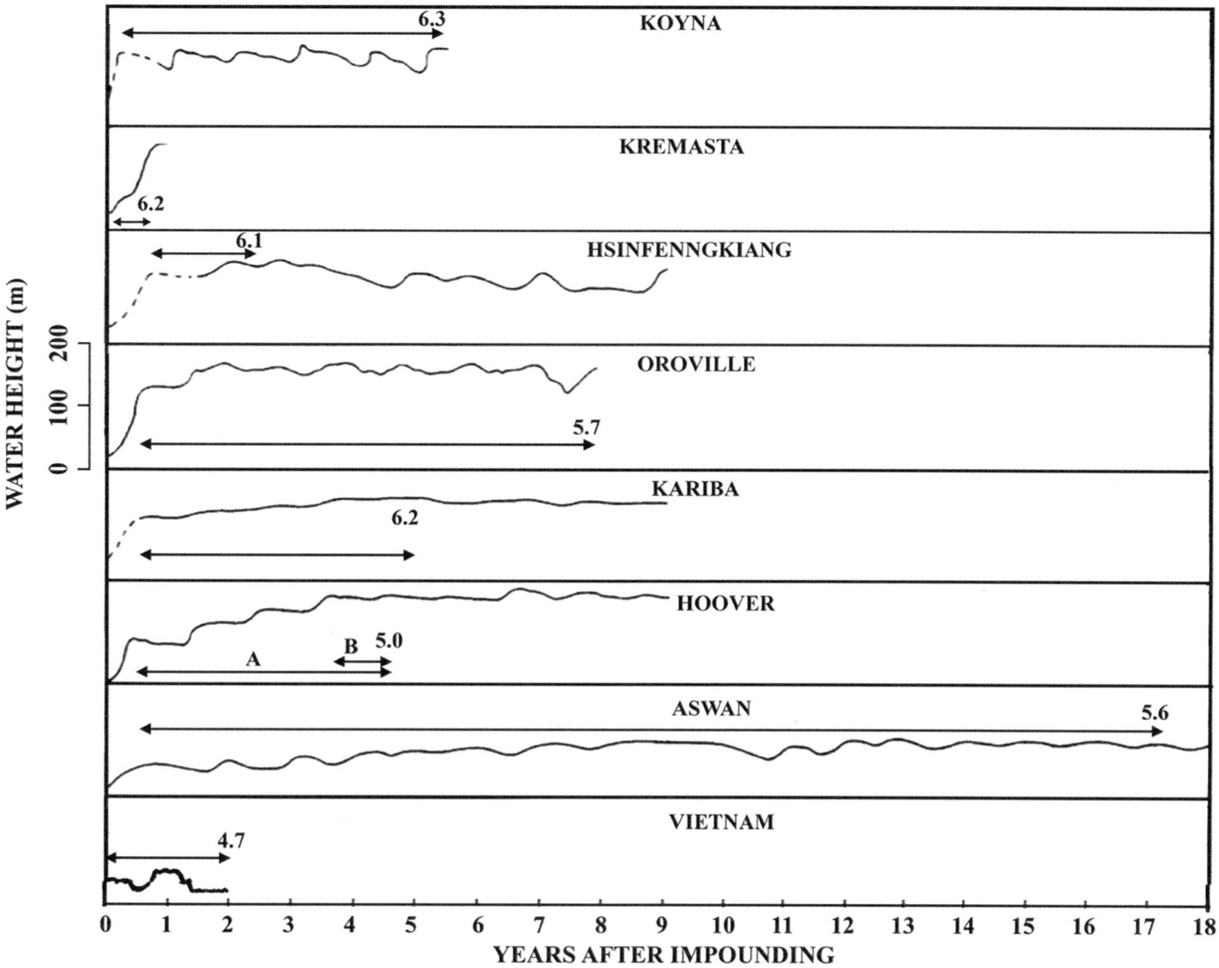

Fig. 2.2 Reservoir levels, largest RIS event and the time interval after which these events occurred for eight major RIS cases. Uncertainties in published data are indicated by dashed lines. Arrows indicate the time intervals. Figure after Gupta (1985); modified from Simpson (1976), and updated from Gupta (1992)

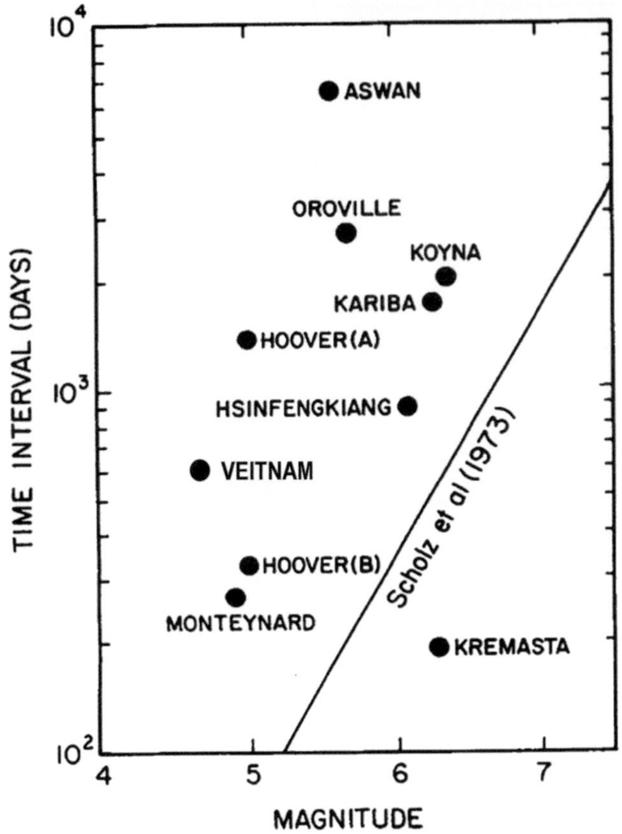

Fig. 2.3 Time interval after first filling of the reservoir plotted against and the largest RTS event that occurred. For Hoover there are two entries: **a** is the time interval after first filling and **b** is time interval between first peak reservoir level reached and occurrence of the largest RTS event. The relation by Scholz et al. (1973) between earthquake magnitude and premonitory phenomenon is also plotted in the diagram. Updated from Gupta (1992)

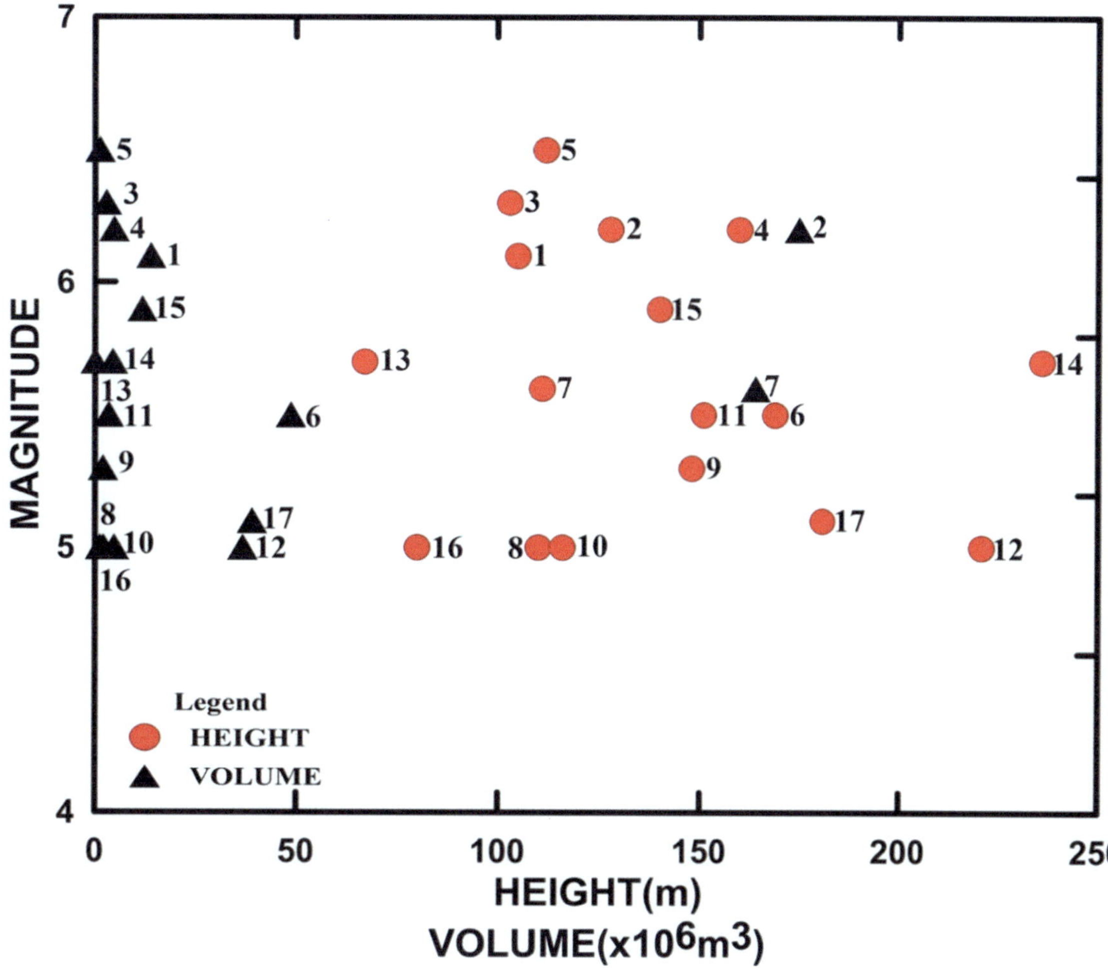

Fig. 2.4 Reservoir volume, height and the magnitude of the largest induced earthquake for seventeen major RIS cases with details provided in Table 2.2. Updated from Gupta (1992)

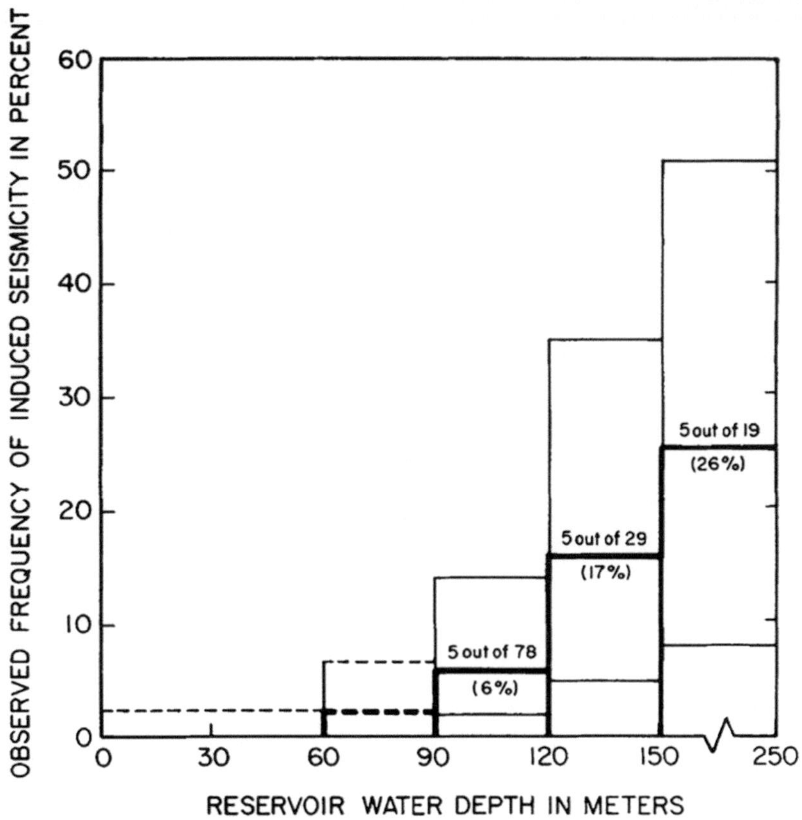

Fig. 2.5 Histogram showing observed frequency percentage (heavy lines) of reservoirs associated with impoundment-triggered seismicity (earthquake magnitudes > 3) within certain water-depth intervals, which percentages are taken as the best estimate of mean probability (includes the three equivocal cases). The light lines are 95% confidence limits of the probability estimates assuming a binomial distribution; the lines are dashed where rough estimates were made for the large number of reservoirs in the lower depth classes. The association of seismicity with impoundment is questionable for two reservoirs considered seismic in the 90–120 m depth category (after Stuart-Alexander and Mark 1976)

Table 2.1 Global distribution of RTS sites where largest event was M 4 and larger (updated from Gupta 2022)

Site number	Name of the dam/ reservoir	Country	Year of impounding	Height of dam (m)	Reservoir volume (10^6 m^3)	Year of the largest earthquake	Magnitude/ intensity	References
Sites where earthquakes having magnitude ≥ 6.0 were triggered								
1	Hsinfengkiang	China (PRC)	1959	105	13,896	1962	6.1	1, 2, 3
2	Kariba Zambia	Zimbabwe	1958	128	175,000	1963	6.2	1, 2, 6
3	Koyna	India	1962	103	2780	1967	6.3	1, 2, 4, 5
4	Kremasta	Greece	1965	160	4750	1966	6.2	1, 2, 4, 5
5	Polyphyto	Greece	1974	112	1220	1995	6.5	37, 38, 39
Sites where earthquakes having magnitude between 5.0 and 5.9 were triggered								
6	Ataturk	Turkey	1990	169	48,700	2017	5.5	40
7	Aswan	Egypt	1964	111	164,000	1981	5.6	2, 7
8	Benmore	New Zealand	1964	110	2040	1966	5.0	1, 2, 8
9	Charvak	Uzbekistan	1971	148	2000	1977	5.3	25
10	Eucumbene	Australia	1957	116	4761	1959	5.0	2
11	Geheyan	China	1993	151	3400	1997	VI	22a
12	Hoover	USA	1935	221	36,703	1939	5.0	1, 2, 10
13	Marathon	Greece	1929	67	41	1938	5.7	1, 2, 4, 5
14	Oroville	USA	1967	236	4400	1975	5.7	2, 11
15	Srinagarind	Thailand	1977	140	11,750	1983	5.9	28
16	Warna	India	1985	80	1260	1993	5.0	24
17	Three Gorges	China	2003	181	39.3	2013	5.1	34
Sites where earthquakes having magnitude between 4.0 and 4.9 were triggered								
18	Aksombo Main	Ghana	1964	134	148,000	1964	V	2, 9
19	Bajina Basta	Yugoslavia	1966	90	340	1967	4.5–5.0	2, 5
20	Bhatsa	India	1981	88	947	1983	4.9	20
21	Bratsk	Russia		100	169	1996	4.2	22b
22	Camarillas	Spain	1960	49	37	1964	4.1	2, 4, 5
23	Canelles	Spain	1960	150	678	1962	4.7	2, 4, 5
24	Capivari–Cachoeira	Brazil		58	180	1971	VI	17
25	Clark Hill	USA	1952	60	3517	1974	4.3	2, 12
26	Dahua	China (PRC)	1982	74.5	420	1993	4.5	26
27	Danjiangkou	China (PRC)	1967	97	16,000	1973	4.7	16
28	Foziling	China (PRC)	1954	74	470	1973	4.5	16
29	Grandwal	France	1959	88	292	1963	V	1, 2, 4, 5
30	Hoa Binh	Vietnam	1988	125		1989	4.9	22c
31	Kastraki	Greece	1968	96	1000	1969	4.6	2
32	Kerr	USA	1958	60	1505	1971	4.9	1, 2, 9
33	Komani	Albania	1985	130	1600	1986	4.2	27
34	Kurobe	Japan	1960	186	149	1961	4.9	2, 13
35	Lake Baikal	Russia					4–4.8a	23
36	Lake Pukaki	New Zealand	1976	106	9000	1978	4.6	21
37	Manicouagan 3	Canada	1975	108	10,423	1975	4.1	2
38	Marimbondo	Brazil	1975	94	6150	1975	IV	18
39	Monteynard	France	1962	155	275	1963	4.9	1, 2, 4, 5
40	Nova Ponte	Brazil	1993	142	12,800	1998	4.0	42

(continued)

Table 2.1 (continued)

Site number	Name of the dam/ reservoir	Country	Year of impounding	Height of dam (m)	Reservoir volume (10^6 m^3)	Year of the largest earthquake	Magnitude/ intensity	References
41	Nurek	Tadjikistan	1972	317	1000	1972	4.6	1, 2, 14
42	P. Colombia/V. Grande	Brazil	1973–1974	40/56	1500/2300	1974	4.2	19
43	Piastra	Italy	1965	93	13	1966	4.4	2, 4, 5
44	Pieve de Cadore	Italy	1949	116	69	1950	V	2, 15
45	Shenwo	China (PRC)	1972	50	540	1974	4.8	16
46	Vouglans	France	1968	130	605	1971	4.4	2, 4, 5
47	Karun-III	Iran	2005	185	2970	2005	4.3	29
48	Itoiz	Spain	2002	111	418	2004	4.6	30
49	Longtan	China	2005	217	7458	2007	4.2	31
50	Dongjing	China	2005	157	755	2010	4.4	32
51	Song Tranh 2	Vietnam	2010	96	733	2012	4.7	33, 41
52	El Cuchillo	Mexico	1993	42	1123	2013	4.5	35
53	Pirris Reservoir	Costa Rica	2011	113	36	2014	4.8	36

References 1 = (Gupta and Rastogi 1976); 2 = (Packer et al. 1979); 3 = (Shen et al. 1974); 4 = (Rothe 1970, 1973); 5 = (Bozovic 1974); 6 = (Gough and Gough 1970); 7 = (Toppozada 1982); 8 = (Adams 1974); 9 = (Simpson 1976); 10 = (Carder 1945); 11 = (Bufe et al. 1976); 12 = (Talwani 1976); 13 = (Hagiwara and Ohtake 1972); 14 = (Soboleva and Mamadaliev 1976); 15 = (Caloi 1970); 16 = (Oike and Ishikawa 1983); 17 = (Berrocal et al. (personal communication) 1989); 18 = (Veloso et al. 1987); 19 = (Berrocal et al. 1984); 20 = (Rastogi et al. 1986); 21 = (Reyners 1988); 22a = (Chen et al. 1996); 22b = (Pavlenov and Sherman 1996); 22c = (Tung 1996); 23 = (Djadkov 1997); 24 = (Rastogi et al. 1997); 25 = (Plotnikova et al. 1992); 26 = (Guang 1995); 27 = (Muco 1991); 28 = (Chung and Liu 1992); 29 = (Kangi and Heidari 2008); 30 = (Jiménez et al. 2009); 31 = (Hua et al. 2013); 32 = (Ma 2012); 33 = (Wiszniowski et al. 2015); 34 = (Zhang et al. 2018); 35 = (Montalvo-Arrieta et al. 2018); 36 = (Ruiz-Barajas et al. 2019); 37 = (Pavlou et al. 2013); 38 = (Pavlou 2019); 39 = (Michas et al. 2020); 40 = (Büyükakpınar et al. 2021); 41 = (Lizurek et al. 2021); 42 = (Raza et al. 2023)

Table 2.2 Table with height of the dam, reservoir volume and the magnitude/intensity of the largest triggered earthquakes

S. No.	Name of the dam/reservoir	Country	Height of dam (m)	Reservoir volume (10^6 m^3)	Magnitude/intensity
1.	Hsinfengkiang	China (PRC)	105	13,896	6.1
2.	Kariba Zambia	Zimbabwe	128	175,000	6.2
3.	Koyna	India	103	2780	6.3
4.	Kremasta	Greece	160	4750	6.2
5.	Polyphyto	Greece	112	1220	6.5
6.	Ataturk	Turkey	169	48,700	5.5
7.	Aswan	Egypt	111	164,000	5.6
8.	Benmore	New Zealand	110	2040	5.0
9.	Charvak	Uzbekistan	148	2000	5.3
10.	Eucumbene	Australia	116	4761	5.0
11.	Geheyan	China	151	3400	VI
12.	Hoover	USA	221	36,703	5.0
13.	Marathon	Greece	67	41	5.7
14.	Oroville	USA	236	4400	5.7
15.	Srinagarind	Thailand	140	11,750	5.9
16.	Warna	India	80	1260	5.0
17.	Three Gorges	China	181	39.3	5.1

References

Adams RD (1974) The effect of Lake Benmore on local earthquakes. Eng Geol 8:155–169

Baecher BG, Keeney RL (1982) Statistical examination of reservoir induced seismicity. Bull Seismol Soc Am 72:553–569

Berrocal J, Assumpcao M, Antezana R, Dias Neto CM, Ortega R, Franca H, Velose J (1984) Sismicidade do Brasil Instituto Astronomico e Geofisico. Universidade de Sao Paulo, Sao Paulo

Berrocal J, Fernandes C, Antezana R, Shukowsky R, Barbosa JR, Shayani S, Pereira ES (1989) Induced seismicity by the Sobradinho reservoir, Bahia, Brazil (personal communication)

Bozovic A (1974) Review and appraisal of case histories related to seismic effects of reservoir impounding. Eng Geol 8:9–27

Bufe CG, Lester FW, Lahr KM, Lahr JC, Seekins LC, Hanks TC (1976) Oroville earthquakes: normal faulting in Sierra Nevada foothills. Science 192:72–74

Büyükakpınar P, Cesca S, Hainz S, Jamalreyhani M, Heimann S, Dahm T (2021) Reservoir-triggered earthquakes around the Atatürk Dam (Southeastern Turkey). Front Earth Sci 9. https://doi.org/10.3389/feart.2021.663385

Caloi P (1970) How nature reacts on human intervention: responsibilities of those who cause and who interpret such reaction. Ann Geofis 23:283–305

Carder DS (1945) Seismic investigations in the Boulder Dam area, 1940–1944, and the influence of reservoir loading on earthquake activity. Bull Seismol Soc Am 35:175–192

Chen B, Li S, Yin Z (1996) On the characteristics and prediction of induced earthquakes of the Geheyan reservoir. In: Abstract volume IASPEI regional assembly in Asia, Tangshan, China, pp 1–3

Chung W-Y, Liu C (1992) The reservoir-associated earthquakes of April 1983 in Western Thailand: source modelling and implications for induced seismicity. PAGEOPH 138(1):17–41

Djadkov PG (1997) Induced seismicity at the Lake Baikal; principal role of load rate. In: Abstract volume IASPEI general assembly, Thessaloniki, Greece, pp 18–28

Gough DI, Gough WI (1970) Load induced earthquakes at Kariba. 2. Geophys J Roy Astron Soc 21:79–101

Guang YH (1995) Seismicity induced by cascade reservoirs, Yantan hydroelectric power stations. In: Proceedings of the international symposium on reservoir-induced seismicity. State Seismological Bureau, Beijing, pp 157–163

Gupta HK (1985) The present status of reservoir induced seismicity investigations with special emphasis on Koyna earthquakes. Tectonophysics 118(3&4):257–279

Gupta HK (1992). Reservoir Induced Earthquakes, Elsevier Scientific Publishing Company, Amsterdam, p 364

Gupta HK (2011) Artificial water reservoir triggered earthquakes. In: Encyclopedia of solid earth geophysics, vol 1. Springer + Business Media, pp 15–24

Gupta HK (2021) Artificial water reservoir triggered earthquakes with special emphasis on Koyna, India. In: Encyclopedia of solid earth geophysics, vol 2, 2nd edn. Springer + Business Media, pp 19–32

Gupta HK (2022) Artificial water reservoir-triggered seismicity (RTS): most prominent anthropogenic seismicity. Surv Geophys. https://doi.org/10.1007/s10712-021-09675-z

Gupta HK, Rastogi BK (1976) Dams and earthquakes. Elsevier, Amsterdam, The Netherlands, p 229

Hagiwara T, Ohtake M (1972) Seismic activity associated with the failing of the reservoir behind Kurobe Dam, Japan, 1963–1970. Tectonophysics 15:241–254

Hua W, Chen Z, Zheng S, Yan C (2013) Reservoir-induced seismicity in the Longtan reservoir, southwestern China. J Seismol 17(2):667–681

Jiménez A, Tiampo KF, Posadas AM, Luzón F, Donner R (2009) Analysis of complex networks associated to seismic clusters near the Itoiz reservoir dam. Eur Phys J Spec Top 174(1):181–195

Kangi A, Heidari N (2008) Reservoir-induced seismicity in Karun III dam (southwestern Iran). J Seismol 12:519–527

Lizurek G, Leptokaropoulos K, Wiszniowski J, Giang NV, Nowaczyńska I, Plesiewicz B, Van DQ, Tymińska A (2021) Seasonal trends and relation to water level of reservoir-triggered seismicity in Song Tranh 2 reservoir, Vietnam. Tectonophysics 820:229121

Ma W (2012) Analysis on the disaster mechanism of rock collapse of M4.4 reservoir-induced earthquake on January 17, 2010, at Dongjing reservoir in Guizhou Province China. Nat Hazards 62(1):141–148

Michas G, Pavlou K, Vallianatos F, Drakatos G (2020) Correlation between seismicity and water level fluctuations in the Polyphyto Dam, North Greece. Pure Appl Geophys. https://doi.org/10.1007/s00024-020-02465-50

Montalvo-Arrieta JC, Pérez-Campos X, Ramos-Zuñiga LG, Paz-Martínez EG, Salinas-Jasso JA, Navarro de León I, Ramírez-Fernández JA (2018) El Cuchillo seismic sequence of October 2013–July 2014 in the Burgos basin, northeastern Mexico: hydraulic fracturing or reservoir-induced seismicity? Bull Seismol Soc Am 108(5B):3092–3106

Muco B (1991) The swarm of Nikaj-Merturi, Albania. Bull Seismol Soc Am 81:1015–1021

Oike K, Ishikawa Y (1983) Induced earthquakes associated with large reservoirs in China. Chin Geophys II 2:383–403

Packer DR, Cluff LS, Knuepfer PL, Withers RJ (1979) A study of reservoir induced seismicity. Woodward-Clyde Consultants, U.S.A. U.S. Geol. Surv. Contract 14-08-0001-16809

Pavlenov VA, Sherman SI (1996) Premises of induced seismicity on the reservoirs of the Angare River. In: Abstract volume IASPEI regional assembly in Asia, Tangshan, China, pp 1–3

Pavlou K (2019) Relationship between observer seismicity and water level fluctuations in Polyphyto Dam area (North Greece). J Geogr Environ Earth Sci Int 21(2):1–10

Pavlou K, Kaviris G, Chousianitis K, Drakatos G, Kouskouna V, Makropoulos K (2013) Seismic hazard assessment in Polyphyto Dam area (NW Greece) and its relation with the "unexpected" earthquake of 13 May 1995 (Ms = 6.5, NW Greece). Nat Hazards 13:141–149

Plotnikova LM, Makhmudova VI, Sigalova OB (1992) Seismicity associated with the Charvak reservoir, Uzbekistan. Pure Appl Geophys 139:607–608

Rastogi BK, Chadha RK, Raju IP (1986) Seismicity near Bhatsa reservoir, Maharashtra, India. Phys Earth Planet Inter 44:179–199

Rastogi BK, Chadha RK, Sarma CSP, Mandal P, Satyanarayana HVS, Raju IP, Kumar N, Satyamurthy C, Nageswara Rao A (1997) Seismicity at Warna reservoir (near Koyna) through 1995. BSSA 87(6):1484–1494

Raza H, Kivi IR, França GS, Vilarrasa V (2023) Reservoir impoundment-triggered seismicity in Brazil: the case of M4.0 Nova Ponte earthquake. Sci Rep 13:22226. https://doi.org/10.1038/s41598-023-48924-6

Reyners M (1988) Reservoir induced seismicity at Lake Pukaki, New Zealand. Geophys J 93:127–135

Rothe JP (1970) The seismic artificials (man-made earthquakes). Tectonophysics 9:215–238

Rothe JP (1973) A geophysics report. In: Ackermann WC, White GF, Worthington EB (eds) Man-made lakes: their problems and environmental effects. Am Geophys Union Geophys Monogr 17:441–454

Ruiz-Barajas S, Santoyo MA, Oterino MB, Alvarado GE, Climent A (2019) Stress transfer patterns and local seismicity related to reservoir water-level variations: a case study in central Costa Rica. Sci Rep 9(1):1–16

Sayão E, França GS, Holanda M, Gonçalves A (2020) Spatial database and website for reservoir-triggered seismicity in Brazil. Nat Hazards Earth Syst 20:2001–2019

Scholtz CH, Sykes LR, Aggarwal YP (1973) Earthquake prediction: a physical basis. Science 181:803–810

Shen C, Chang C, Chen H, Li T, Hueng L, Wang T, Yang C, Lo H (1974) Earthquakes induced by reservoir impounding and their effect on the Hsinfengkiang Dam. Sci Sin 17(2):239–272

Simpson DW (1976) Seismicity changes associated with reservoir loading. Eng Geol 10:123–150

Soboleva OV, Mamadaliev UA (1976) The influence of the Nurek Reservoir on local earthquake activity. Eng Geol 10:293–305

Stuart-Alexander DE, Mark RK (1976) Impoundment-induced seismicity associated with large reservoirs. U.S. Geol. Surv. Open File Rep. 76-770

Talwani P (1976) Earthquakes associated with Clark Hill Reservoir, South Carolina: a case of induced seismicity. Paper presented at the 1st international symposium on induced seismicity. Eng Geol 10:239–253

Toppozada TR (1982) UNDP/Tokten report on Aswan earthquakes

Tung NT (1996) The induced seismicity at Hoa Binh Reservoir region. In: Abstract volume IASPEI regional assembly in Asia, Tangshan, China, pp 1–3

Veloso JAV, Assumpcao M, Concalves ES, Reis JC, Duarte VM, da Motta CBG (1987) Registro de sismicidade induzida em reservatorios da CEMIG e FURNAS: an 50 Congr. Bras Geol Eng 1:135–146

Wilson MP, Foulger GR, Gluyas JG, Davies RJ, Julian BR (2017) HiQuake: the human-induced earthquake database. Seismol Res Lett 88(6):1560–1565

Wiszniowski J, Van Giang N, Plesiewicz B, Lizurek G, Van DQ, Lasocki S (2015) Preliminary results of anthropogenic seismicity monitoring in the region of Song Tranh 2 reservoir Central Vietnam. Acta Geophys 63(3):843–862

Zhang L, Li J, Sun X, Liao W, Zhao Y, Wei G, He C (2018) A possible mechanism of reservoir-induced earthquakes in the Three Gorges Reservoir, Central China. Bull Seismol Soc Am 108(5B):3016–3028. https://doi.org/10.1785/0120180015

3 Characteristics of RTS

Gupta et al. (1972a, b) analyzed the characteristics of the available RTS sequences and compared them with the characteristics of normal earthquake sequences occurring in the nearby areas with similar geological background. The common characteristics identified for the RTS sequences differed considerably from the normal earthquake sequences. Here, we briefly discuss these characteristics.

3.1 Frequency-Magnitude Relationship

During 1930s, two Japanese seismologists, M. Ishimoto and K. Iida conducted outstanding research relating the frequency of earthquakes with their magnitude (Ishimoto and Iida 1939). After studying several local and regional earthquake sequences, they established a simple relation between earthquake frequency and magnitude for a given region:

$$\text{Log}\, N = A - b\text{M},$$

where N is the number of shocks of magnitude $\geq$ M and A and b are constants specific for the region. Basically, the value of A is based on the duration of observation, the size of the region under consideration and b is governed by the ratio of the number of earthquakes in low to high magnitude range. Utsu (1965) carried out detailed observations on the earthquake magnitude and frequency in several regions and came up with the following relation for empirically calculating b values:

$$b = \frac{0.4343m}{\sum_{i=1}^{m} M_i - mM_{\min}}$$

In the above relation m is the total number of earthquakes and $M_{\min}$ is the lowest magnitude considered. A global value of 0.5–1.5 was observed for b by Isacks and Oliver (1964). They also observed that for most cases the b value was between 0.7 and 1.0. Gupta et al. (1972b) noted that for the RTS earthquake sequences, globally, the observed b values for the foreshocks were comparable with the aftershocks b values and both were higher than the regional b values. This observation was found to hold good for the RTS cases reported globally in the subsequent years (Gupta 2021, 2022).

An excellent example of changes in the b value consequent to creation of an artificial water reservoir has been provided by Simpson and Negmatullaev (1981) for the Nurek Reservoir in erstwhile U.S.S.R. Figure 3.1 taken form Gupta (1985) depicts the reservoir area within the rectangle. Simpson and Negmatullaev (1981) reported that the b values outside the reservoir area have remained unchanged for the last several decades. Inside the rectangular area, the b values changed from 0.8 to 1.0 consequent to the impounding of the reservoir in 1973.

H. K. Gupta, *Artificial Water Reservoir Triggered Seismicity*, https://doi.org/10.1007/978-3-032-05873-7_3

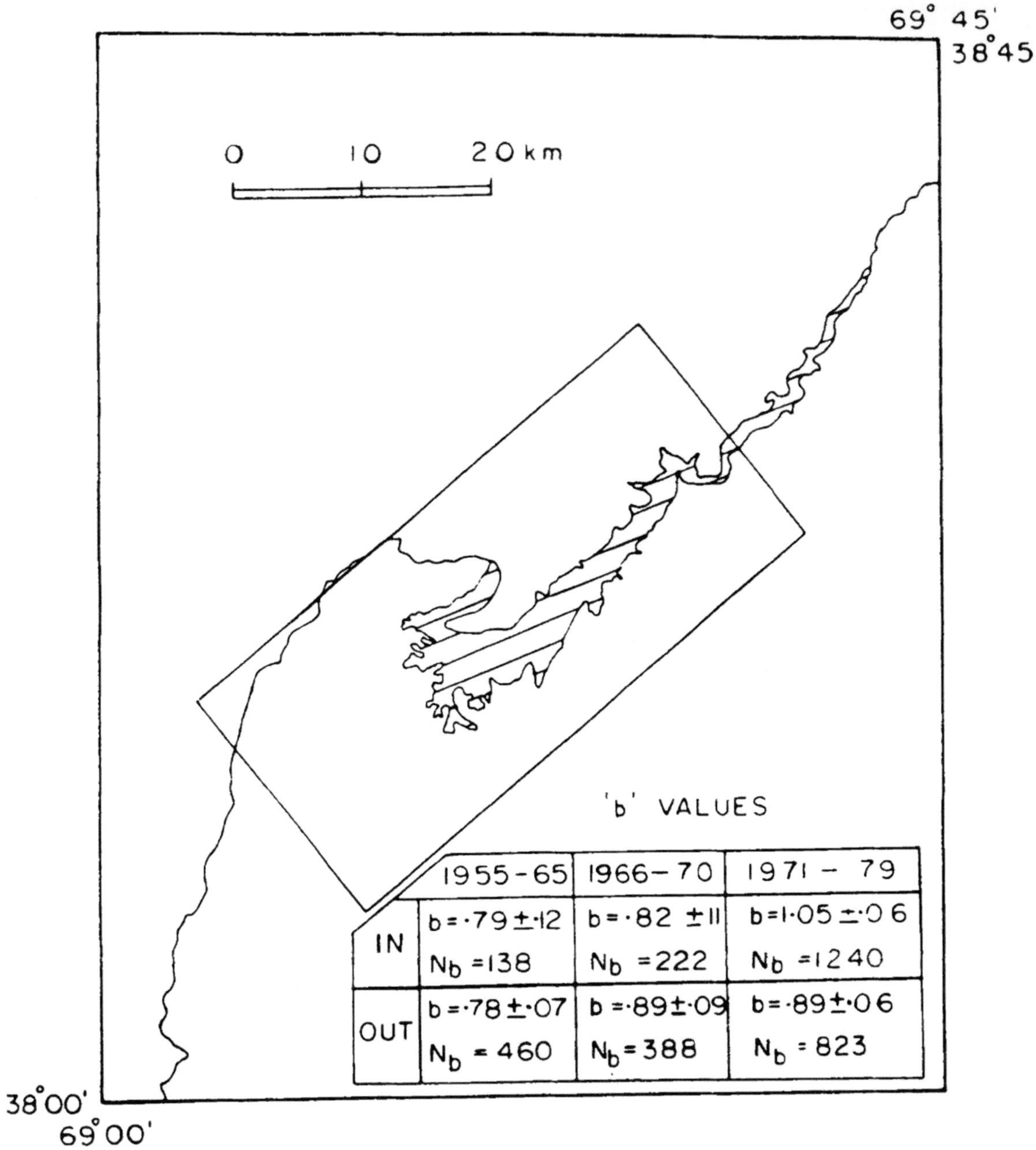

Fig. 3.1 Figure based on data from Simpson and Negmatullaev (1981) and adapted from Gupta (1985). The area within the rectangle is referred to as the reservoir area. The b values "outside" the reservoir area have more or less remained unchanged over the last several years, while the b values "inside" the reservoir area have increased considerably consequent to filling of the reservoir

3.2 The Magnitude Ratio of the Largest Aftershock to the Main-Shock

In 1965, the following relation was established between the largest aftershock magnitude (M_1) and the main shock magnitude (M_0) by Bath (1965):

$$M_0 - M_1 = 1.2$$

The validity of this relation was confirmed by a study of over 200 earthquake sequences by Papazachos (1971). For the RTS sequences, Gupta et al. (1972a) found the relation of the largest aftershock to the mainshock magnitude to be ~ 0.9. Later, Papazachos (1974), from a study of RTS sequences established the following relation with a standard deviation of 0.3:

$$M_0 - M_1 = 0.6$$

Overall, it has been observed that RTS sequences are characterized by high b values and the ratio of the largest aftershock to the main shock being also high compared to the normal earthquake sequences in the nearby regions (Table 3.1).

3.3 The Decay of Aftershocks

Utsu (1961) developed an inverse power law relation for the distribution of the aftershocks of an earthquake:

$$n(t) = Ct^{-h}$$

In the above relation, $n(t)$ is the number of aftershocks per unit of time, C and h are constants and t is the time elapsed after the main shock. The rate of decay of the aftershocks is indicated by h. It is governed by the physical and stress state of the aftershock zone, as explained by Mogi (1962).

Over the past 60 years, this relation has been used to infer the state of stress distribution and its release in the aftershock earthquake zone. Shen et al. (1974) pointed out that the decay of the aftershock activity of the Hsinfengkiang RTS event of M 6.1 in 1962 was much slower compared to other earthquakes occurring in the nearby region. Similar observations were made by Morrison et al. (1976) for decay of Oroville RTS event of M 5.7 of 1975. In the years to follow, several similar observations were reported. Gupta (1992) listed aftershock durations of RTS events for 10 RTS sites. In the years to follow, many more similar observations have been made (Table 3.2).

3.4 Patterns of Foreshock-Aftershock Occurrence

After preforming detailed laboratory and field investigations, Kiyoo Mogi broadly divided the foreshock-aftershock patterns of earthquake sequences into three types (Mogi 1963a, b). These are:

Table 3.1 The magnitudes of the main shock and largest aftershock and the *b* values for reservoir-associated earthquake sequences

Region	Main shock magnitude (M_0)	Largest after shock magnitude (M_1)	$M_0 - M_1$	M_1/M_0	b	Source[a]
Lake Mead	5.0	4.4	0.6	0.88	1.40	1
Monteynard	4.9	4.5	0.4	0.92	0.72	1
Kariba	6.1	6.0	0.1	0.98	1.02	1
Kremasta	6.2	5.5	0.7	0.89	1.12	1
Koyna	6.3	5.1	0.8	0.83	1.09	1
Hsinfengkiang	6.1	5.3	0.8	0.87	1.04	2
Oroville	5.7	5.1	0.6	0.89	0.61	3
Nurek	4.6	4.3	0.3	0.93	1.05	4
Aswan	5.6	4.6	1.0	0.82		5
Bhatsa	4.9	3.9	1.0	0.80	1.0	6[b]
Ataturk	5.5	5.1	0.4	0.92	0.9	7
Itoiz	4.6	3.8	0.8	0.82	0.67	8
Three Georges	5.1	4.6	0.5	0.90	0.97	9

[a] 1: Gupta and Rastogi (1976); 2: Shen et al. (1974); 3: Morrison et al. (1976); 4: Simpson and Negmatullaev (1981); 5: Kebeasy et al. (1987); 6: Patil et al. (1986); 7: Büyükakpınar et al. (2021); 8: Ruiz et al. (2006); 9: Yao et al. (2017)

[b] Patil et al. (1986) have depicted variation in b values from 1 July 1983 through 30 Sept, 1983; *b* value for the entire sequence is estimated from an arithmetic average of peak and trough values in the *b* value curve

Table 3.2 The relations showing the decay of aftershock activity

Region	$n(t) = Ct^{-h}$	Unit of time (days)	Total time	Source[a]
Kariba	$130t^{-1.0}$	1	60 days	1
Kremasta	$134t^{-0.78}$	1	200 days	2
Koyna	$180t^{-1.0}$	1	110 days	1
Koyna	$1342t^{-0.77}$	15	Dec. 1967–Dec. 1971	3
Oroville	$112t^{-0.78}$	1	100 days	4
Aswan	$20t^{-1.0}$	1	100 days	5
Bhatsa	$300t^{-1.0}$	1	500 days	6

[a] 1: Gupta et al. (1972b); 2: Comninakis et al. (1968); 3: Guha et al. (1974); 4: Rajendran and Gupta (1986); 5: Kebeasy et al. (1987); 6: Rastogi et al. (1986)

Type 1. When the Earth strata are homogeneous and stress is uniformly applied, there would be no foreshocks, and the main shock would be followed by numerous aftershocks.
Type 2. When Earth strata are rather heterogeneous and stresses applied are not uniform, small foreshocks would occur preceding the main shock. Aftershocks shall follow in a relatively large number.
Type 3. When the Earth strata are extremely heterogeneous and applied stresses have varying concentration, a swarm of earthquakes would occur where the number would increase gradually peaking and then decaying. There shall be no main shock.

These are depicted in Fig. 3.2I.

Conducting a global review of RTS till 1970s, Gupta et al. (1972a, b) suggested that the above-mentioned characteristics of earthquake sequences reflect the mechanical properties of the rock media where the earthquakes are occurring. The deviation from the normal implies that filling of the artificial water reservoirs has made a change introducing heterogeneity in the media. This can be visualized in the cartoon shown in Fig. 3.2II, where 'A' represents a homogeneous media rock volume. In this situation when the stress exceeds the strength of the rock, the main earthquake would occur releasing most of the strain. There would be small peripheral aftershocks. For such an earthquake sequence, there would be no foreshock, the largest aftershock would be much smaller than the mainshock and the aftershock activity would be over soon. So, the '*b*' value for this sequence would be small, the ratio of the largest aftershock to the mainshock would be small and the aftershocks would not last for long. This is typically the case with Type 1 of Mogi's models mentioned above. Such earthquake sequences are normally observed in stable continental regions. If a reservoir is built and the percolation of water introduces heterogeneity in the media and fragmentation of the competent rock would result ('B' in Fig. 3.2II). The accumulated strain would be released as and when the competence of an individual fragmented rock mass is exceeded. The largest of the fragmented rock volume would become the main shock. The competence of certain fragmented rock masses may be exceeded before the largest rock volume, causing the foreshocks. There would also be several fragments whose competence would be subsequently exceeded giving rise to the aftershocks. In such a case the '*b*' value of the sequence would be high, the ratio of the largest aftershock to the mainshock magnitude would also be high and the aftershocks would continue for a longer time compared to Type 1. This is typically the Type 2 of Mogi's classification of earthquake sequences. And when the rock volume is extremely fragmented, there would be several fragments of similar size. In such a situation there would be no main shock. There would be several shocks of similar magnitude and that would constitute an earthquake swarm, belonging to Type 3 of Mogi's models.

A deadly M 6.2 earthquake, named the Latur earthquake occurred on 29 September 1992 with epicenter located 300 km northeast of Koyna (Fig. 3.3). It claimed over 11,000 human lives (Gupta 1993). It was also located in the Deccan Volcanic Province (DVP) of India, like the location of the Koyna 1967 earthquake. The characteristics of the Latur earthquake sequence fell in Type 1 of Mogi's model. The ratio of the largest aftershock to the mainshock was 0.6; the *b* value was 0.36; there were no foreshocks, and aftershock did not last long. This was quite in contrast with the 1967 Koyna earthquake sequence which had a *b* value of 1.09 (Gupta 1992) and the ratio of the largest aftershock to the main shock was 0.83, and the aftershocks continued for a long time. It may also be mentioned that the Latur earthquake sequence had characteristics like Peninsular Shield of India which is characterized with a *b* value of 0.47.

The above discussed characteristics of RTS sequences which differ from the normal earthquake sequences of the concerned region were discovered in 1972 (Gupta et al 1972a, b). Over the years the RTS sequences, globally, are found to have characteristics belonging to Type II of Mogi's model (Gupta 2022), while normal earthquake sequences in the nearby region fall in Type 1 of Mogi's models.

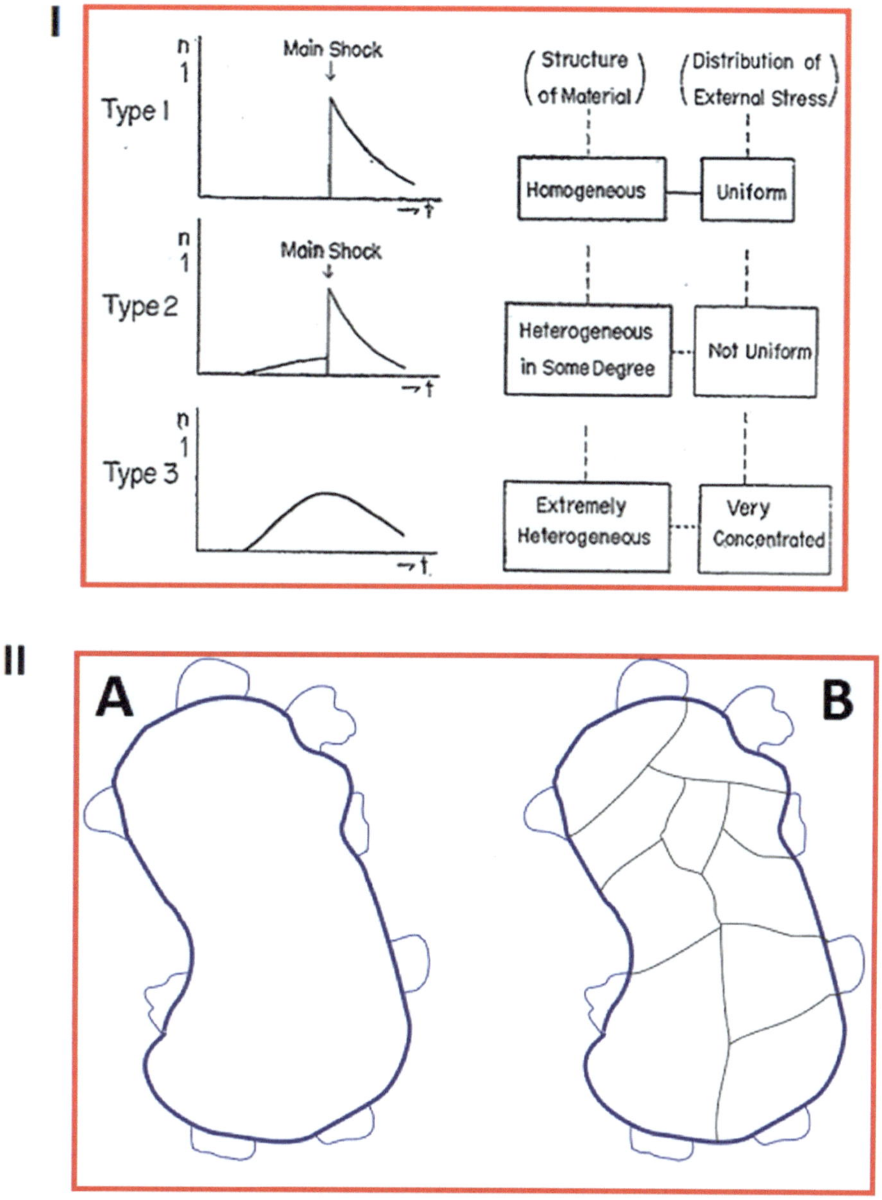

Fig. 3.2 **I** Depicts (Mogi 1963a, b) classification of the earthquake sequences into three broad categories. **II** Cartoon for (A) homogenous rock mass and (B) fragmented rock mass. For details see the text

3.5 Palghar Earthquake Swarm

We have discussed the Type 2 of Mogi's (1963a, b) earthquake sequence: the earthquakes occurring in Koyna is a classic example. The Latur M 6.2 earthquake belonged to Type 1. Recently, during 2018–2020, the Deccan Volcanic Province (DVP) of South India also hosted the Type 3 of Mogi's model earthquake sequence with the occurrence of Palghar earthquake swarm. The Palghar region (Fig. 3.3) is situated on a thick pile of Deccan Basalt varying in thickness from 1.4 to 1.7 km (Kaila et al. 1981a, b). The earthquake swarm it hosted had 4854 earthquakes in the 0.1–4.1 (M_L) magnitude range, and focal depths 4 and 16 km (Srinagesh et al. 2020). The swarm covered an area of 10 km × 6 km. One of the causative factors for this swarm suggested by Sharma et al. (2020) is intense rain fall preceding the swarm of earthquakes. Pavankumar et al. (2020) suggested that the presence of upper crustal heterogeneities coupled with basement fault and low rheological strength coupled with fractured upper and middle crust in the region may have been responsible for the observed swarm.

Gupta (2022) noted that the 1967 Koyna M 6.3 earthquake was preceded by a noticeable foreshock and aftershock activity. While the Latur M 6.2 earthquake on 29

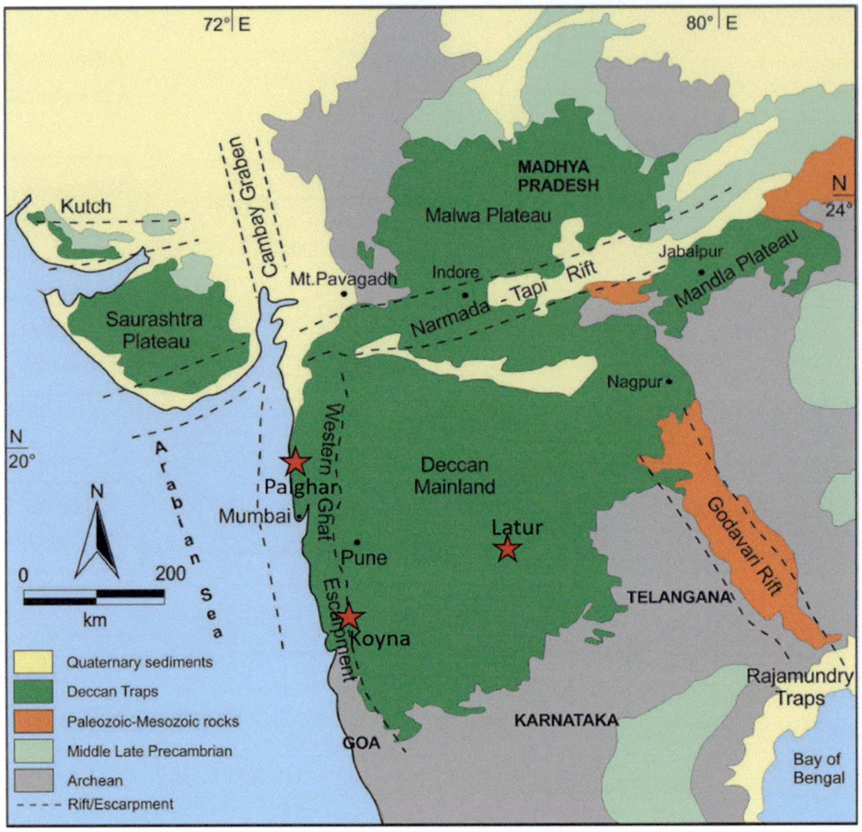

Fig. 3.3 Locations of the Koyna earthquake of 10 December 1967 of M 6.3 and the Latur earthquake of 29 September 1993 of M 6.2 and the Palghar swarm (2018–2020). All three are located in the Deccan Volcanic Province (DVP) of India. For details, see the text

Table 3.3 b-values of earthquake sequences in the Deccan Volcanic Provinces and Peninsular India (after Gupta 2022)

Region and time window	Number of earthquakes	Magnitude range	b-value
Koyna reservoir aftershocks 10-12-1967–27-6-1969 (1)	422	3.0–5.4	1.09
Latur aftershocks (2)	161	≥ 3.0	0.44
30-09-1993–20-01-1994	43	≥ 1.5	0.36
Palghar (3)	8766	0.4–4.7	2.5 (modelled)
Indian Peninsula earthquakes from historical times to more than 300 years (1)	52	4.0–7.0	0.47

References 1 = (Gupta and Rastogi 1976); 2 = (Rastogi 1994); 3 = (Mandal et al. 2021).

September 1993 had no foreshocks, and the aftershocks did not last for too long. Whereas the Palghar swarm of 2018–2020 did not have any mainshock-aftershock kind of situation (Mandal et al. 2021). It is therefore inferred that on the bases of foreshock-aftershock pattern, while the RTS events at Koyna belong to Type 2, the Latur earthquake sequence belong to Type 1 and Palghar swarm to Type 3 of Mogi's model. Table 3.3 provides an overview of the three types of earthquake sequences in DVP of India.

The Deccan Volcanic Province of India have the distinction of hosting all the three kinds of earthquake sequences as classified by Mogi (1963a, b).

References

Bath M (1965) Lateral inhomogeneities in the upper mantle. Tectonophysics 2:483–514

Büyükakpınar P, Cesca S, Hainzl S, Jamalreyhani M, Heimann S, Dahm T (2021) Reservoir-triggered earthquakes around the Atatürk Dam (southeastern Turkey). Front Earth Sci 9:1–18

Comninakis P, Drakopoulos J, Moimioulidis G, Papazachos BC (1968) Foreshock sequences of the Kremasta earthquake and their relation to the water loading of the Kremasta artificial lake. Ann Geofis (Rome) 21:39–71

Guha SK, Gosavi ED, Nand K, Eadale JG, Marwadi SC (1974) Koyna earthquakes (Oct. 1963 to Dec. 1973). Central Water and Power Research Station, Khadakwasla, Poona, 24 pp

Gupta HK (1985) The present status of Reservoir Induced Seismicity investigations with special emphasis on Koyna earthquakes. Tectonophysics 118:257–279

Gupta HK (1992) Reservoir Induced Earthquakes, Elsevier Scientific Publishing Company, Amsterdam, p 364

Gupta HK (1993) The deadly Latur earthquake. Science 262:1666–1667

Gupta HK (2021) Artificial water reservoir triggered earthquakes with special emphasis on Koyna, India. In: Encyclopedia of solid earth geophysics, vol 2, 2nd edn. Springer + Business Media, pp 19–32

Gupta HK (2022) Artificial water reservoir-triggered seismicity (RTS): most prominent anthropogenic seismicity. Surv Geophys. https://doi.org/10.1007/s10712-021-09675-z

Gupta HK, Rastogi BK (1976) Dams and earthquakes. Elsevier, Amsterdam, p 229

Gupta HK, Rastogi BK, Narain H (1972a) Common features of reservoir associated seismic activities. Bull Seismol Soc Am 62:481–492

Gupta HK, Rastogi BK, Narain H (1972b) Some discriminatory characteristics of earthquakes near Kariba, Kremasta, Koyna artificial lakes. Bull Seismol Soc Am 63:493–507

Isacks B, Oliver J (1964) Seismic waves with frequencies from 1 to 100 cycles per second recorded in a deep mine in northern New Jersey. Bull Seismol Soc Am 54:1941–1979

Ishimoto M, Iida K (1939) Observations sur lés sesmes enregistrés par le microsismographe construit dernièrement. Bull Earthq Res Inst 17:443–478

Kaila KL, Murthy PRK, Rao VK, Kharetchko GE (1981a) Crustal structure from deep seismic soundings along the Koyna II (Kelsi-Loni) profile in the Deccan Trap area, India. Tectonophysics 73(1981):365–384

Kaila KL, Reddy PR, Dixit MM, Lazarenko MA (1981b) Deep crustal structure at Koyna, Maharashtra, indicated by deep seismic sounding. J Geol Soc India 22(1981):1–16

Kebeasy RM, Maamoun M, Ibrahim EM (1987) Aswan Lake induced earthquakes. Bull USEE 19:155–160

Mandal P, Srinagesh D, Suresh G, Naresh B, Naidu M, Singh DK, Swathi K, Devi A, Vittal S, Vengala PK, Vijaya Raghavan R, Shekar M, Saha S (2021) Characterization of earthquake hazard at the Palghar and Pulichintala swarm activity regions (India) through three-dimensional modelling of b-value and fractal (correlation) dimensions. Nat Hazards 108:1183–1196

Mogi K (1962) Study of elastic shocks caused by the fracture of heterogeneous materials and its relation to earthquake phenomena. Bull Earthq Res Inst 40:125–173

Mogi K (1963a) The fracture of a semi-infinite body caused by an inner stress origin and its relation to the earthquake phenomena (second paper). The case of the materials having some heterogeneous structures. Bull Earthq Res Inst 41:595–614

Mogi K (1963b) Some discussions on aftershocks, foreshocks and earthquake swarms—the fracture of a semi-infinite body caused by an inner stress origin and its relation to the earthquake phenomena (third paper). Bull Earthq Res Inst 41:615–658

Morrison PW, Stump BW, Uhrhammer R (1976) The Oroville earthquake sequence of August 1975. Bull Seism Soc Am 66:1065–1080

Papazachos BC (1971) Aftershock activity and aftershock risk in the area of Greece. Ann Geofis (Rome) 24:439–456

Papazachos BC (1974) On the relation between certain artificial lakes and the associated seismic sequences. Paper presented at international colloquium on seismic effects of reservoir impounding, Mar 1973. Royal Society London. Eng Geol 8:39–48

Patil DN, Bhosale VN, Guha SK, Powar KB (1986) Reservoir induced seismicity in the vicinity of Lake Bhatsa, Maharashtra, India. Phys Earth Planet Inter 44:73–81

Pavankumar G, Chakravarthi NN, DemuduBabu M, Manglik A (2020) Magnetotelluric study of an intraplate seismic zone in the Palghar region of the Deccan Volcanic Province India. J Earth Syst Sci 129(2020):1–16

Rajendran K, Gupta HK (1986) Was the earthquake sequence of August 1975 in the vicinity of Laker Oroville, California, reservoir induced? Phys Earth Planet Inter 44:142–148

Rastogi BK (1994) Latur earthquake: not triggered. Mem Geol Soc India 35(1994):131–138

Rastogi BK, Chadha RK, Raju LP (1986) Seismicity near Bhatsa reservoir, Maharashtra, India. Phys Earth Planet Inter 44:179–199

Ruiz M, Gaspà O, Gallart J, Díaz J, Pulgar JA, García-Sansegundo J, López-Fernández C, González-Cortina JM (2006) Aftershocks series monitoring of the September 18, 2004, M = 4.6 earthquake at the western Pyrenees: a case of reservoir-triggered seismicity? Tectonophysics 424:223–243

Sharma V, Wadhawan M, Rana N, Sreejith KM, Agrawal R, Kamra C, Hosalikar KS, Narkhede K, Suresh G, Gahalaut VK (2020) A long duration non-volcanic earthquake sequence in the stable continental region of India: the Palghar swarm. Tectonophysics 779(2020):1–7

Shen C, Chang C, Chen H, Li T, Hueng L, Wang T, Yang C, Lo H (1974) Earthquakes induced by reservoir impounding and their effect on the Hsinfengkiang Dam. Sci Sin 17(2):239–272

Simpson DW, Negmatullaev SK (1981) Induced seismicity at Nurek Reservoir, Tadjikistan, USSR. Bull Seismol Soc Am 71(5):1561–1586

Srinagesh D, Singh DK, Vikas G, Naresh B, Roy S, Mahesh G, Murty YVVBSN, Raju PS, Suresh G, Mandal P, Sharma AN, Sekhar M, Tiwari VM (2020) An appraisal of recent earthquake activity in Palghar region. Curr Sci 110:1592–1598

Utsu T (1961) A statistical study on the occurrence of aftershocks. Geophys Mag (Tokyo) 30(40):523–605

Utsu T (1965) A method for determining the value of 6 in the formula log n = α − 6M, showing the magnitude-frequency relation for earthquakes. Geophys Bull Hokkaido Univ 13:99–103

Yao Y-S, Wang Q-L, Liao W-L, Zhang L-F, Chen J-H, Li J-G, Yuan L, Zhao Y-N (2017) Influences of the Three Gorges Project on seismic activities in the reservoir area. Sci Bull 62:1089–1098

4 Globally a Few Important RTS Sites

4.1 Lake Mead, USA

Lake Mead, located in Nevada-Arizona region of USA, is a classic example of RTS where way back in 1945, Dean S. Carder of USGS provided scientific evidence of association of earthquakes with filling of the Lake Mead (Carder 1945). The 132 m high Hoover Dam created Lake Mead. It is located west of the Colorado Plateau in the southeast margin of Basin and Range province. Starting from where Colorado River flows out of Grand Canyons, Lake Mead flows westward for about 80 km. Lake Mead is crossed by several north–south trending mountain ranges. A good description of the local geology around Lake Mead has been provided by Longwell (1936, 1963), Anderson (1971), Rogers and Gallanthine (1974), among others. Large scale volcanic activity occurred in the region, beginning in Early Tertiary that continued to Cenozoic. Five episodes of intense volcanic activity have been documented. At places, in an around the Lake Mead, thickness of the volcanic pile extends to 5 km.

Impounding of Lake Mead commenced in 1935. As reported by Jones (1944), and Raphael (1954), no earthquakes were reported in the vicinity of Lake Mead. Filling of Lake Mead reached the peak in 1937 which was followed with the occurrence of several tremors in the vicinity of the lake (Carder 1945). As can be noted from Fig. 4.1, in the subsequent years the peak of water levels in Lake Mead were followed by an increase in tremor frequency in the vicinity of the reservoir. The seismic activity in the vicinity of Lake Mead peaked in 1939 with the occurrence of a M 5 earthquake on 11 June 1939. Another M 5 earthquake occurred on 9 September 1942. Recently, in 2014, lake Mead again got filled to the capacity and that was followed by the occurrence of some 20 tremors of M ~ 2 (Beal 2014). With a capacity of 367 km^3, Lake Mead is the largest man-made water reservoir in the US. However, with continuous drought in region over the past couple of decades, the water levels in Lake Mead have reached the lowest level since its impoundment in 1935.

Based on Gupta (2022).

4.2 Hsinfengkiang, China

The 105 m high Hsinfengkiang Dam is created on the east–west trending huge Late Mesozoic granitic mass, located 160 km northeast of Canton in Guangdong Province of China. The 13,896 million m^3 capacity reservoir created by the Hsinfengkiang Dam is surrounded by Upper Paleozoic-to Jurassic and Cretaceous strata in the north and south, forming an extensive mountain area. To the east of the reservoir lies a northeast trending long and narrow formations of Tertiary era. Volcanic eruptions during Upper Jurassic and basalt lava flows of Early Quaternary age occurred in the region (Shen et al. 1974). Prior the impounding of the reservoir, there are only four mild earthquakes reported from the region during the preceding 25 years of the impoundment. Filling of the reservoir started in October 1959, which was followed with the occurrence of tremors in the region. These were recorded by the seismological observatory at Canton. This necessitated setting up of a local network of seismic stations to record the reported tremors. Originally the dam was constructed to withstand tremors of VI Intensity on the Modified Mercalli (MM) scale. However, due to reported initiation of tremors following the impoundment of the reservoir, the dam was strengthened to withstand intensity VIII earthquakes on the MM Intensity scale. The site was rocked by a M 6.1 earthquake on 19 March 1962. RTS at Hsinfengkiang has been continuously discussed (Shen et al. 1974; Wang et al. 1976; Hsu et al. 1976; Zhu et al. 1982; Gupta 2022). It has been found that the strongest earthquake of M 6.1 occurred in the vicinity of the dam where the water depth was maximum. Most of the hypocenters are

H. K. Gupta, *Artificial Water Reservoir Triggered Seismicity*, https://doi.org/10.1007/978-3-032-05873-7_4

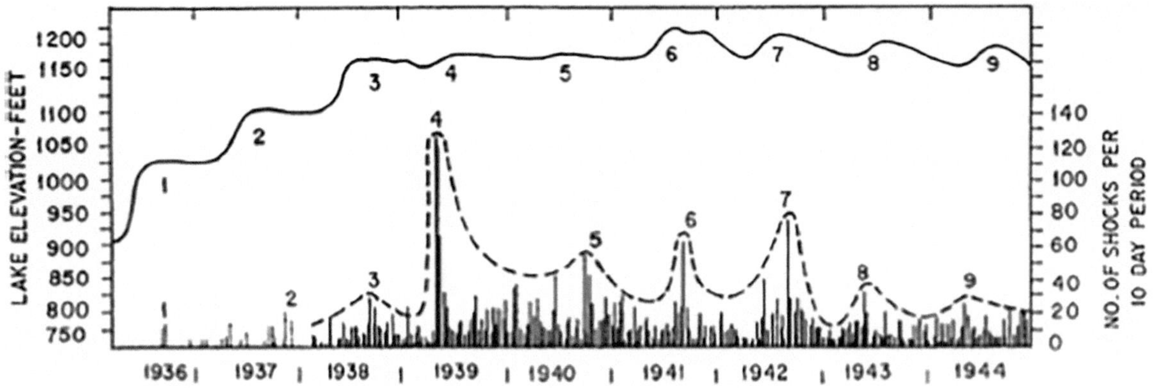

Fig. 4.1 Water levels and local seismicity at Lake Mead, USA. Only felt reports are plotted for the years 1936 and 1937. The rises in water levels and the corresponding bursts of seismic activity are numbered. General trend of tremor-frequency variation is shown by dotted lines (after Carder 1945)

in the vicinity of the intersection of the faults at shallow depths of 1–11 km. There was a remarkable increase in the frequency and magnitude of the tremors within one year of the impoundment of the reservoir. The M 6.1 earthquake occurred soon after the peak level was reached (Fig. 4.2). The main shock was followed by strong aftershocks of M 5.1. There are several magnitudes 4 and larger earthquakes occurring in the following years. However, the level of 1962 seismic activity remained unmatched.

Shen et al. (1974) have discussed the foreshocks and the aftershocks occurring in the vicinity of Hsinfengkiang Dam in detail. They report that there was almost a three-fold increase in the number of tremors in the two months preceding the occurrence of the 19 March 1962 M 6.1 magnitude mainshock. This enhanced seismic activity continued till about three weeks before the main shock. During the three weeks preceding the mainshock no M $\geq$ 3 foreshock was recorded. Shen et al. (1974) report that over 80,000 foreshocks were recorded in the 28 months period following the impoundment of the lake and occurrence of the 19 March 1962 M 6.1 earthquake. It has been also pointed out that while the foreshocks focal depths were mostly shallower that 4 km, the aftershocks had focal depths up to 7 km.

Shen et al. (1974) have discussed the characteristics of Hsinfengkiang RTS sequence. Several similarities are found with the reporting of the common characteristics of RTS sequences deciphered by Gupta et al. (1972a, b). These include:

1. Intense foreshock activity occurred preceding M 6.1 earthquake on 19 March 1962. The mainshock was followed with aftershocks of up to magnitude 5.3.
2. The *b* value for foreshocks was 1.12; and for aftershocks 1.02. Both these values are higher than the regional *b* value of 0.72.
3. The largest aftershock to the main shock magnitude ratio was 0.87.
4. The aftershocks attenuated rather slowly.

Gupta (1992) studied all the reports of Hsinfengkiang RTS and inferred the following:

1. While the foreshocks mostly occurred near the dam site and farther northwest of the reservoir, the aftershocks concentrated near the dam and northeast of it.
2. The aftershocks occurred mostly away from the dam in contrast to the foreshock's activity.
3. 90% of tremors occurred within 5 km of the reservoir periphery.
4. With the passage of time, while the foreshocks were shallower (3–6 km), the aftershocks had deeper focal depth.
5. The main shock of 19 March 1962 had a very well-defined foreshock-aftershock pattern.

From the above, it may be inferred that the initial seismic activity and the 19 March 1962 mainshock were triggered due to increased elastic shear stress of the water load causing an increase in the pore pressure because of compaction, while the later aftershocks could be attributed pore pressure diffusion (Bell and Nur 1978; Simpson 1986). In a review of RTS in China, Chen and Talwani (1999) reported that 313 M $\geq$ 3, and 49 M $\geq$ 4 triggered earthquakes have been reported from Hsinfengkiang in China.

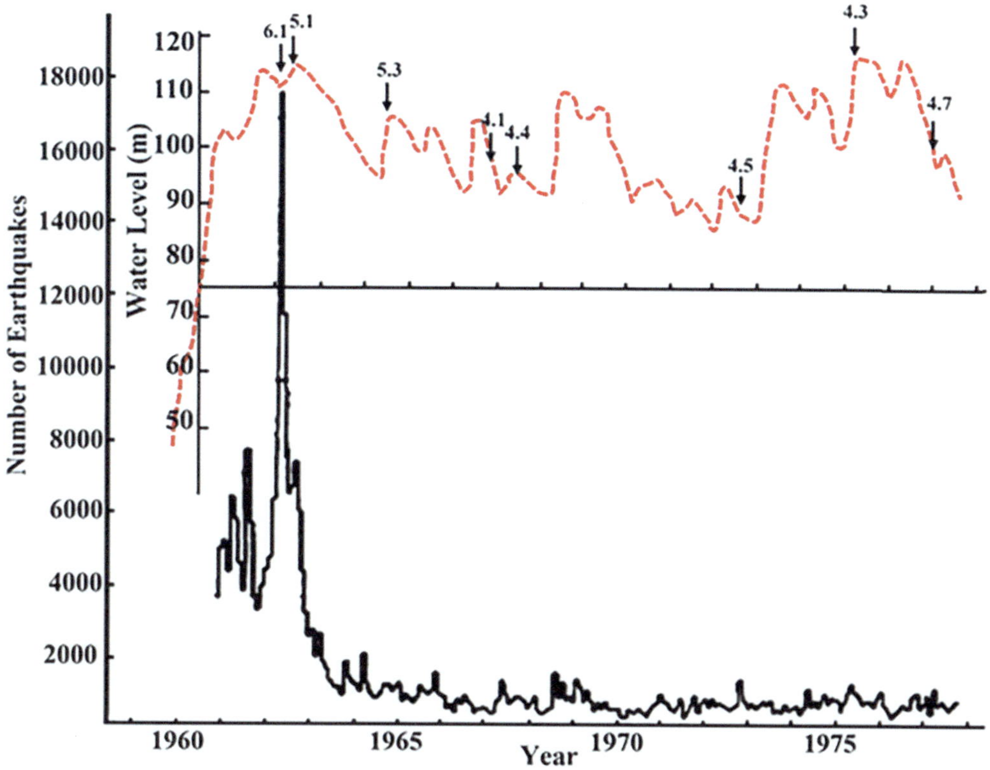

Fig. 4.2 Water levels in the Hsinfengkiang reservoir, China, and monthly frequency of earthquakes. Earthquakes of M $\geq$ 4 are marked by arrows (modified from Ishikawa and Oike 1982)

4.3 Kariba Dam, Zambia-Zimbabwe Border

Kariba Dam, located at the border of Zimbabwe-Zambia, has created one of the largest artificial water reservoirs, globally. The 617-m long, 128-m tall Kariba Dam holds 175 million km^3 of water. Lake Kariba is over 223 km long and up to 40 km wide. The depth of the reservoir varies from 29 to 97 m. By volume, it is world's largest man-made reservoir. It is almost four times bigger than the Three Gorges Reservoir in China. The reservoir encompasses several small islands. Zambia is located on the northwest side and Zimbabwe on the southeast side of the reservoir. Karoo sediments of Carboniferous to Triassic age lie in the basement of the reservoir. Precambrian basement is exposed in several locations of the reservoir because of weathering. The basement is characterized with fractures and faults. A good description of the geology of the area is provided by Bond (1953, 1960), Hitchon (1958), and Knill and Jones (1965). In review, Snow (1974) has provided an overview of the geological aspects of the region related with hydrology, impoundment of the reservoir and consequent seismicity.

There was no seismic station operating in the vicinity of Lake Kariba before its impoundment. Hence, no authentic information about the earthquakes occurring in the region is available. There is geological evidence of earthquake occurrence along a fault at Binga near the west end of the lake (Archer and Allen 1969). Seismic stations were installed in the vicinity of the lake, including a WWSSN at Bulawayo in 1963. Over 2000 tremors were recorded by these stations. Most of these were in the rift valley of Mid-Zambezi hosting the Lake Kariba. Figure 4.3 depicts the seismicity and the reservoir levels at Lake Kariba from 1959 to 1972. An increase in seismicity is observed beginning in February 1962 that ended in November 1962. There was another burst of seismic activity coinciding with the peak water levels in the reservoir. The strongest earthquake of M 6.2 occurred on 23 September 1963. At times an increase in seismic activity is also observable coinciding with the rapid falloff the reservoir level. Annual fluctuation of water levels at Kariba are of the order of 2.5 m. An increase in seismic activity is also observable during September to November 1971 including an M 3.8 earthquake.

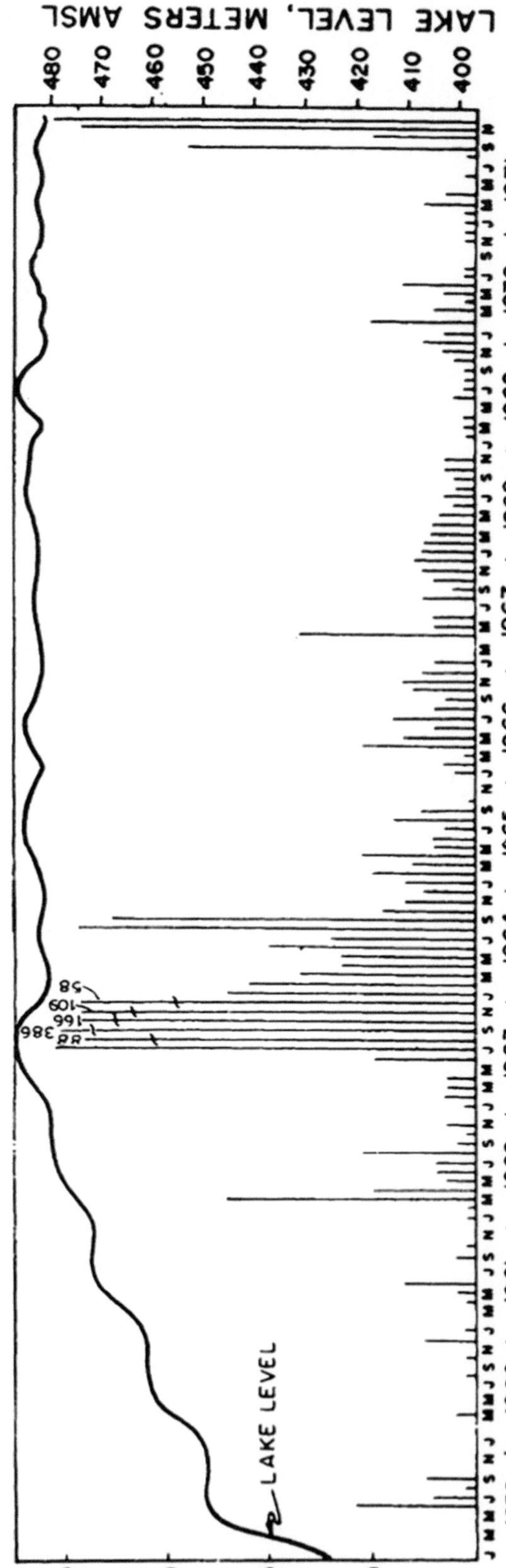

Fig. 4.3 Monthly tremor frequency and water levels of Lake Kariba

After a gap several years tremors of M 4 occurred on 1st and 2nd April 2020 in the vicinity of Kariba Dam. These were attributed to the fluctuations in water levels in the Kariba Dam by the Zambezi River Authority (ZRA). An M 4.9 earthquake occurred on the 27 December 2024 in the vicinity of Lake Kariba. It has been attributed to water level fluctuations in the lake ZRA. This earthquake was preceded by M 4.1 to M 4.7 earthquakes on 1 December 2024.

It may be concluded that Kariba is a rare RTS site which is found to be active after several decades of initiation of RTS consequent to the impoundment of the reservoir way back in 1959.

4.4 Kremasta, Greece

Kremasta Dam site located in Greece is among the half a dozen RTS sites globally where earthquakes of M $\geq$ 6 have occurred. A reservoir of 4.75 million km^3 is created by the 147 m tall Kremasta Dam built across Achelous River in Greece. Snow (1972) has summarized the geology of the region. The area in the vicinity of Kremasta Dam has been divided by thrust faults into the Pindus, Gavrovo and Ionion zones. In the east of Kremasta lies the Pindus zone constituted by silicious limestones of Jurassic and Cretaceous age with ophiolites, shales and cherts followed by flysch deposits. In the west of Kremasta is the Ionian zone which also has neritic limestones. Kremasta itself lies in the Gavrovo zone dominated by massive, gray, neritic limestones.

Kremasta is in seismically active region of Greece. However, according to Galanopoulos (1967), there was no significant activity in the region during the period extending from 1700 to 1965. In the area where the river Achelous flows, no magnitude 6 earthquake had been reported since 1821 till the beginning of filling of the reservoir in 1965. A rapid loading of the reservoir started in November 1965. This rapid loading was followed with the increase in seismic activity culminating in the occurrence of a M 6.2 earthquake on 5 February 1966 (Fig. 4.4). According to Stein et al. (1982), the observed seismic activity in the vicinity of the Kremasta Dam could be divided in two significant different segments, firstly, the 5 February 1966 mainshock followed by an intense aftershock activity, characterized by predominantly normal faulting on a NW-striking, SW-dipping fault plane. The second segment consists of the late October activity associated with another main event with thrust fault movement on a plane different from the 5 February 1966 earthquake. Stein et al. (1982) observe that Kremasta is in a transition zone between extensional and compressional tectonics. They conclude that while the 5 February 1966 event was reservoir triggered, the same could not be said for the later event.

4.5 Koyna, India[1]

Koyna Dam, located near the west coast of India in a 65-Ma-old flood basalt province, is known to be the most prominent RTS site globally. The 807.2-m-long, 103.2-m-tall Koyna Dam is built on the Koyna River that flows due south for about 60 km cutting 600-m-deep valley through the Deccan Traps, and then, it turns sharply toward the east and joins Krishna River after some 60 km. The Koyna Dam was built just before the Koyna River takes a sharp east turn, creating the Shivaji Sagar Lake that holds 2780 million m^3 water. In this area, two distinct geomorphic trends are observed: the high plateau of Sahyadri Mountains limited on the west by the continental divide that forms a magnificent scarp and gently rolling country almost at the sea level. The geology of the region has been described in detail by Krishnan (1960), Geological Survey of India (1968) and the Committee of Experts (1968) and several others.

Soon after the impoundment of the reservoir, earthquakes began to occur. The scientifically accepted largest and most damaging triggered earthquake of M 6.3 occurred on 10 December 1967. It claimed over 200 human lives, and the Koyna Township was in shambles (Narain and Gupta 1968). Talking to elderly residents, it was revealed that they had never felt earthquakes before the impoundment of the reservoir in 1962. Tremors became prevalent soon after the impoundment in 1962. The 10 December 1967 earthquake was preceded by a widely felt M 5.5 earthquake on 13 September 1967 (Guha et al. 1970). There is a seismic station operating some 110 km away from Koyna at Pune. Examination of records for several years before 1962 showed no occurrence of tremors in the region. Another water reservoir, Warna, located ~ 30-km SSE of Koyna was impounded in 1985. This further enhanced the occurrence of triggered earthquakes in the region. For convenience, the Koyna and Warna reservoir region is referred to as Koyna region. The seismic activity in the region has continued with 22 M ~ 5 earthquakes, some 200 M ~ 4 earthquakes and several thousand smaller earthquakes (Gupta 2017, 2018).

The Koyna region experienced major bursts of seismic activity during 1967, 1973, 1980, 1993–1994 and 2005. The correspondence between RTS at Koyna and factors like the rate at which the reservoir is loaded, what are the highest water levels reached and what is the duration of the retention of the high-water levels has been well established (Gupta et al. 1972a, b). It was also pointed out that a rate of loading of 12 m/week is a necessary but not a sufficient condition for the M ~ 5 earthquakes to occur in the Koyna region (Gupta 1983).

[1] Chapters 6 and 7 of the book deal with the detailed work carried out in Koyna during the periods 1962–2004 and 2004–2024.

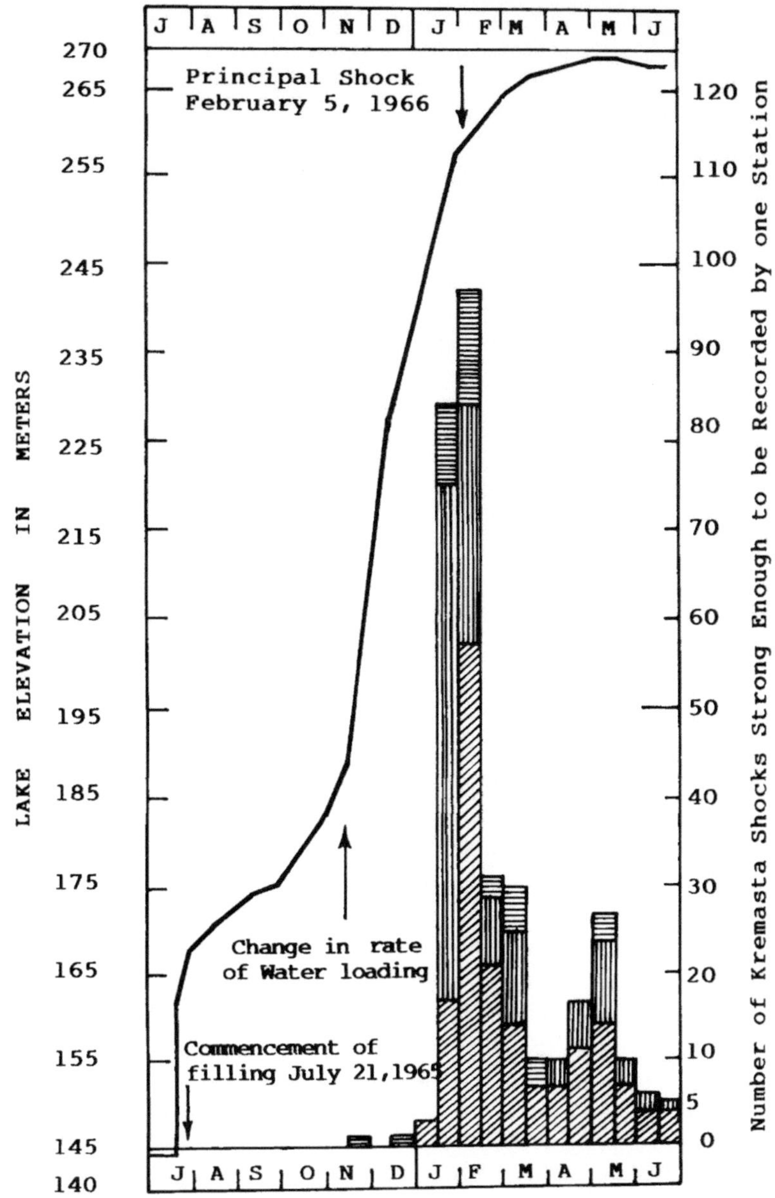

Fig. 4.4 Lake Kremasta elevations and earthquake frequency in the region (after Galanopoulos 1967). Number of Kremasta shocks strong enough to be recorded by: horizontal rules = one station, vertical rules = two stations, slanted rules = several stations per 15-day period

4.6 Polyphyto Dam, Greece

Northern Greece was rocked by a M 6.6 earthquake on the 13 May 1995. It was the largest earthquake to have hit Greece in the past 10 years. The cities of Kozani and Grevena and nearby villages were severely damaged (Hatzfeld et al. 1995). This earthquake occurred in a region where no such earthquake had been reported for the past several hundred years. ISC catalogue also confirm the nonoccurrence of such an earthquake in the region. Papazachos and Papazachou (1989) have documented occurrence of a destructive earthquake in the region in February 896 AD which had destroyed Veria city. The spatiotemporal seismic activity in the vicinity of Aliakmonas River in the northwest Greece has been studied in detail by Pavlou (2019) for the period 1974–2010. The occurrence of the 13 May 1995 earthquake necessitated such an enquiry as the region was known not to be seismically active. The epicenter of the 13 May 1995 earthquake was located 18 km from the southern edge of Polyphyto Lake. The changes in seismic hazard of the region in the vicinity of Polyphyto Lake were addressed earlier by Pavlou et al. (2016). Among the three dams built on the Aliakmonas River, Polyphyto is the biggest with a height of 112 m creating a reservoir where the maximum water depth is 95 m, and the capacity of the reservoir is 1220 million m^3. Polyphyto was first filled in January 1974. Filling of the reservoir initiated RTS in the region which culminated in the M 6.6 earthquake on 13 May 1995. Filling of two other smaller dams, Asomata and Sfikia began on 10 October 1984 and 13 March 1985 respectively. As reported Pavlou et al. (2013), earthquakes of M 5.4 and 5.2 occurred on 10 October 1984 and 13 March 1985 close to these two recently constructed dams. Overall, enhanced region seismicity was witnessed in the region because of the filling of the lakes created by the construction of these three dams. The characteristics of the RTS sequences witnessed at these three reservoirs have been investigated by Pavlou (2019). These are found to be like those, identified by Gupta et al. (1972a, b). Another dam, the Ilarion Dam located upstream of Polyphyto Lake, was completed and impounded on 13 July 2012. Two earthquakes of M 4.6 and M 4.7 occurred on 2 and 3 July 2013 (Pavlou personal communication, 2021; Michas et al. 2020). The M 4.6 earthquake of 14 July 2019 located some 10 km from the Polyphyto Lake is the latest significant event in the region.

The correlation in the water level fluctuations and earthquake occurrences in the vicinity of Polyphyto lake has been investigated by Michas et al. (2020). The seismicity during the period 1996 through 2017 was cross correlated with the water level fluctuations. Moderate size earthquakes (M ~ 4) were found to occur after a gap of 1–2 months from the time of reaching the peak levels in the reservoir. This observation is consistent with the hypothesis of pore-pressure diffusion mechanism for RTS.

4.7 Was the 2008 Mw 7.8 Sichuan, China Earthquake Triggered by Zipingpu Reservoir?

A devastating earthquake of Mw 7.9 occurred on the 12 May 2008, in the western Sichuan region, rupturing the Longmen Shan Fault for some 300 km (Lin et al. 2009) in China claiming some 70,000 human lives and causing unprecedented damage to civil structures. Figure 4.5 depicts the

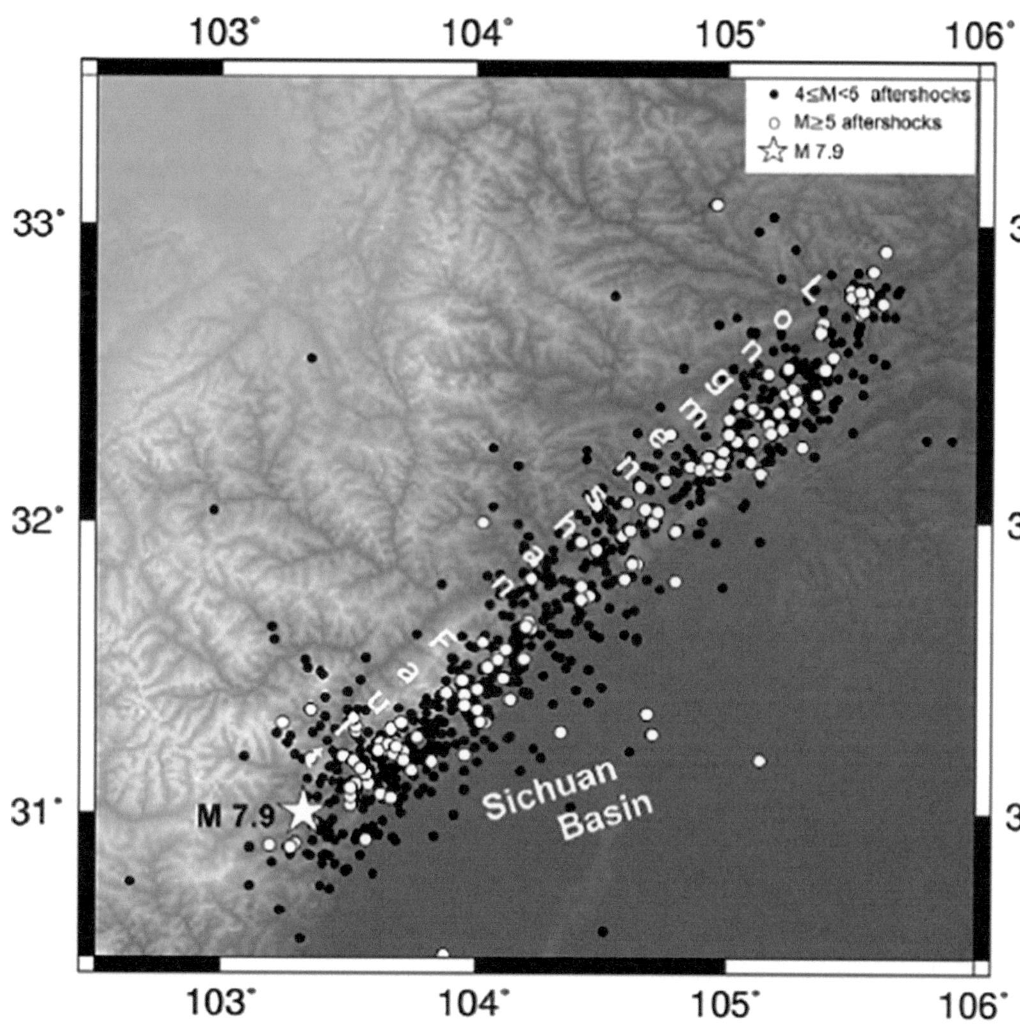

Fig. 4.5 Location of the 12 May 2008 Mw 7.9 Sichuan earthquake (white color star) and M ≥ 5 earthquakes (white filled dots), and 4 ≤ Mw < 5 (black filled dots) till May 2010 (after Gupta and Rekapalli 2022)

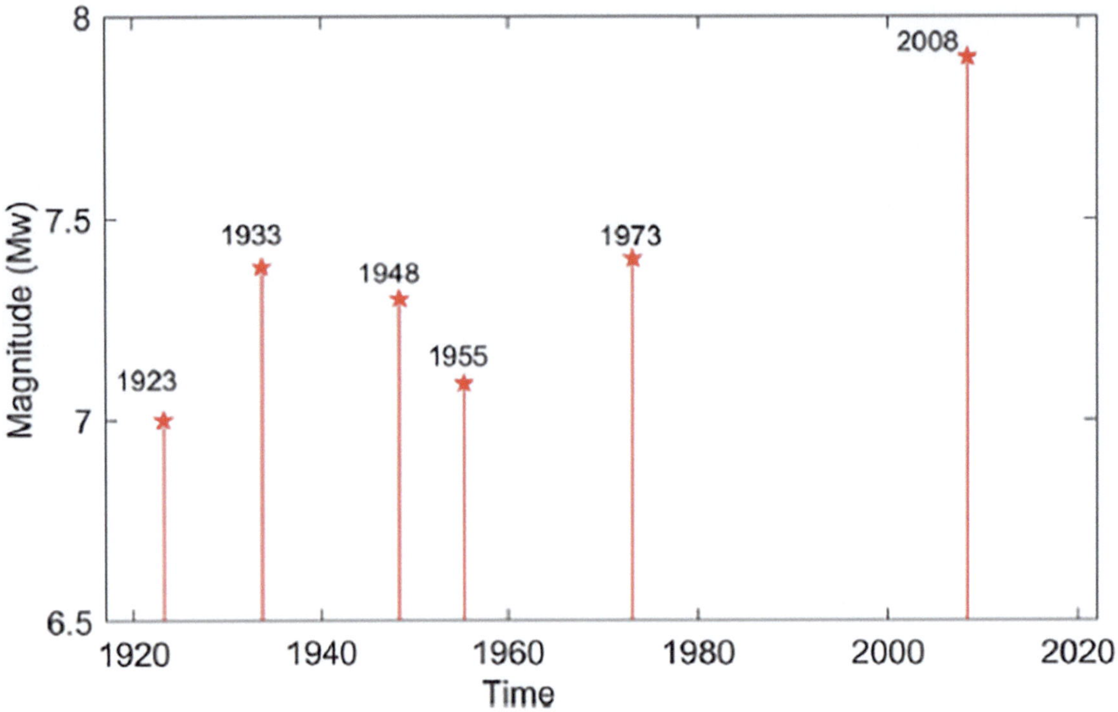

Fig. 4.6 Earthquakes of Mw ≥ 7.0 that occurred within a 300 km radius of the epicenter of the 12 May 2008 Sichuan earthquake of Mw 7.9 since 1920 (after Gupta and Rekapalli 2022)

epicenter of the Sichuan earthquake and its Mw ≥ 4 aftershocks till May 2010 (after Gupta and Rekapalli 2022). The region has a history of major earthquakes exceeding magnitude Mw 7 (Chen and Hsu 2013; Gupta et al. 2008; Gupta and Rekapalli 2022). Figure 4.6 depicts Mw ≥ 7 earthquakes that occurred within 300 km of the epicenter of Sichuan earthquake since 1920. This figure demonstrates that the region has a history of major earthquakes occurring in the past.

Wieland (2017) has addressed the question of the Sichuan earthquake and its being triggered by the nearby Zipingpu Reservoir. He observes that prior to the Sichuan earthquake of 2008, only a few large dams in China had been exposed to large ground motions generated by earthquakes. He observes that the issue of Sichuan earthquake could have been triggered due to the impoundment of the Zipingpu which is based on the fact that large reservoirs are globally known to have triggered earthquakes and moreover the epicenter of Sichuan earthquake is located just 17 km from the Zipingpu reservoir. Moreover, a small branch of the Zipingpu reservoir intersects the fault that ruptured during the Sichuan earthquake.

A joint session of the International Commission on Large Dams (ICOLD) and Chinese National Committee on Large Dams was organized by Wieland (2017) during 29 March–4 April 2009. Over a dozen global experts and over a dozen Chinese experts discussed the issue of the Sichuan earthquake having been triggered by the Zipingpu reservoir (Wieland 2010). A few important statements are copied below:

> The maximum water level of the Zipingpu reservoir (elevation 875.4 m asl) has not exceeded the natural water level (elevation 877 m asl) where the Min River is crossing the Beichuan-Yinxiu fault. Therefore, the original hydrogeological conditions of the Beichuan-Yinxiu fault have not been affected by the impounding of the Zipingpu reservoir.
>
> In August 2004 (Fig. 4.7), 13 months before reservoir impounding, an earthquake monitoring network with seven fixed stations was set up in the reservoir region. In October 2006, the reservoir water level reached elevation 875.4 m asl, 1.6 m below normal storage level. The corresponding reservoir volume was 900 million m^3. In July 2006 and May 2007, the water level was lowered to elevation 820 m asl. Before the Wenchuan earthquake (April 30, 2008), the reservoir water level was at elevation 828.95 m asl, the corresponding reservoir volume was 300 million m^3. From the statistics of the annual seismic activity of the reservoir obtained from the Department of Reservoir Earthquake Research of the Sichuan Seismological Bureau, the frequency and intensity of seismic activity after impounding of the Zipingpu reservoir was almost unchanged before and after reservoir impounding (Fig. 4.7). The recorded seismicity has also no relation with the reservoir water level variation.

Wieland (2017) further observes that Sichuan earthquake has no characteristics of a triggered earthquake, and there is no factual evidence of its being triggered by Zipingpu reservoir.

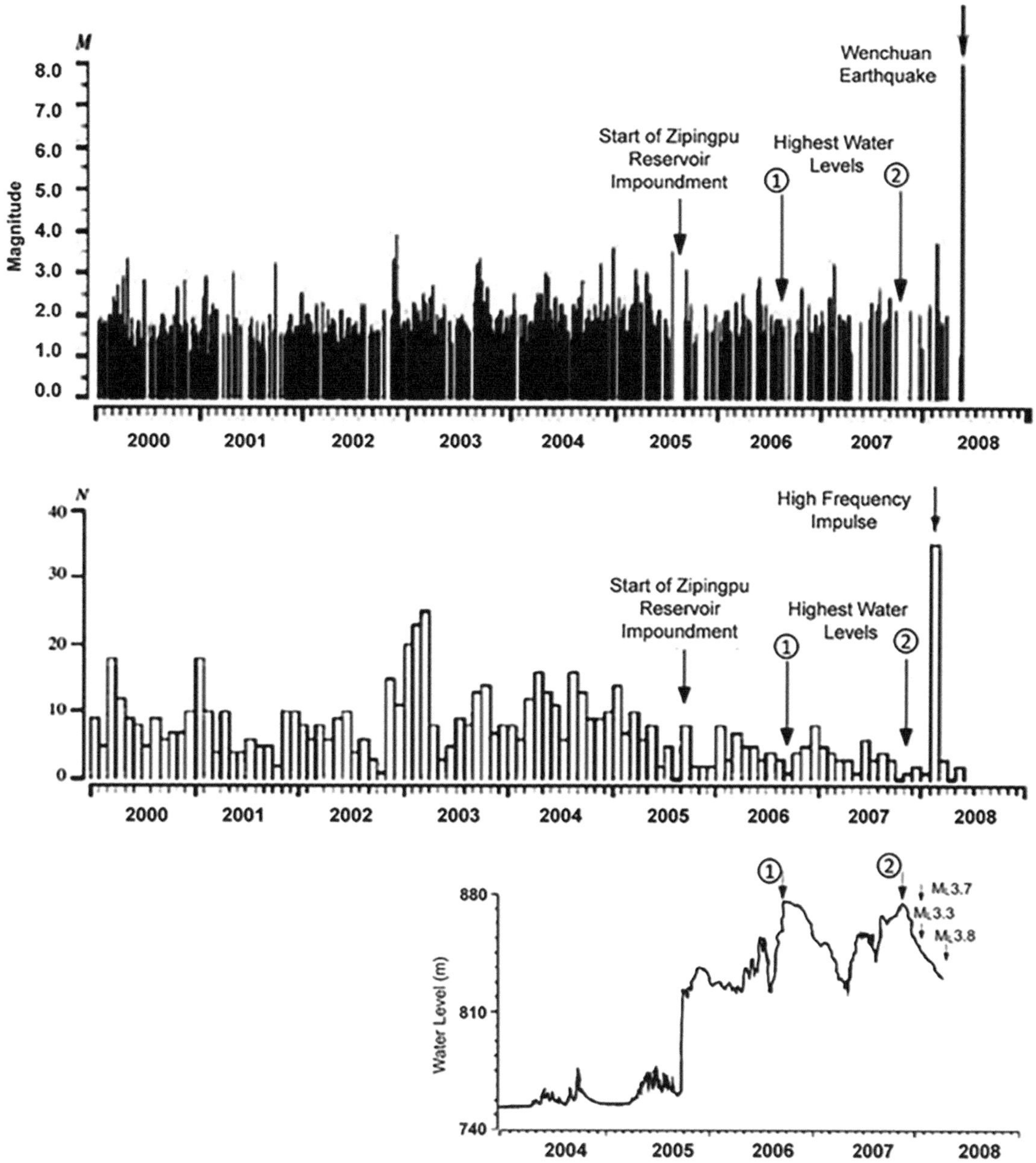

Fig. 4.7 Distribution of magnitude and frequency in the region of Zipingpu reservoir. **a** Magnitude distribution (ML $\geq$ 1.0), **b** frequency distribution (ML $\geq$ 1.0), and **c** reservoir level fluctuation (after Wieland 2017)

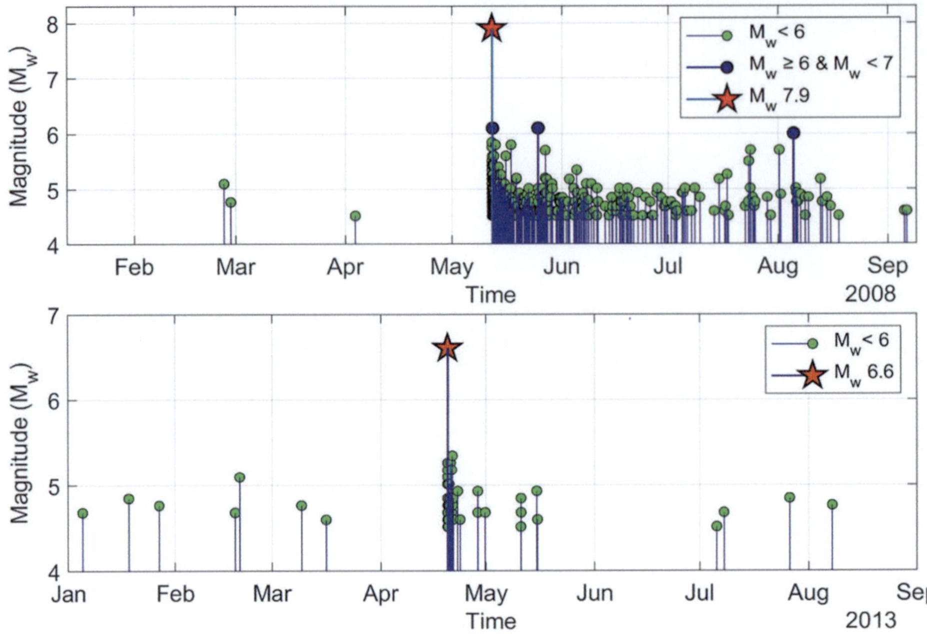

Fig. 4.8 Top: Earthquake occurrence for 241 days (the main Sichuan earthquake of 12 May 2008 occurred on the 121st day). Bottom: Mw 6.8 earthquake of 20 April 2013. The foreshocks and aftershocks depicted occurred within 300 and 150 km radius of the mainshocks of 12 May 2008 and 20 April 2013. For details see the text

Recently (Gupta and Rekapalli 2025) examined the Sichuan earthquake sequence to see whether it has the RTS characteristics. Figure 4.8 shows the Sichuan earthquake sequence for the period 13 January 2008 through 9 September 2008. They found that the Sichuan earthquake sequence had only one foreshock of $M \geq 5$ before the 12 May 2008 main earthquake. The earthquake was followed by an intense aftershock activity and many $M \geq 4$ aftershocks (Fig. 4.8). The 'b' value for the foreshocks is 0.47 and aftershock 1.15 (Fig. 4.9). All these characteristics belong to Type 1 of Mogi's models (Mogi 1963). Another earthquake of Mw 6.8 occurred on 20 April 2013 in the nearby region. This earthquake is also characterized by Type 1 of Mogi's (1963) model.

An additional important fact is that the Sichuan earthquake occurred in a thrust fault tectonic setting (Gahalaut and Gahalaut 2010). It has been shown that the RTS does not occur in the thrust fault tectonic setting. As a matter of fact, Gupta and Rajendran (1986) had examined the seismicity in the vicinity of 11 large reservoirs (exceeding the capacity of 1 km^3 and/or the height of water column being more than 100 m) and found no RTS occurring in the vicinity of these large reservoirs in the Himalayan region. The non-occurrence or RTS had been inferred due the trust fault prevalent tectonic set up which is not conducive for RTS.

The question of Mw 7.9 Sichuan earthquake on the 12 May 2008, being triggered by the Zipingpu reservoir continues to be debated (Chen et al. 2010; Zhang 2011; Tao et al. 2015, and others). However, it may be concluded that the earthquake sequence was very different from RTS sequences and occurred in trust fault tectonic setting which is not conducive for RTS. It may be concluded that the Zipingpu reservoir had no role in triggering the Sichuan earthquake.

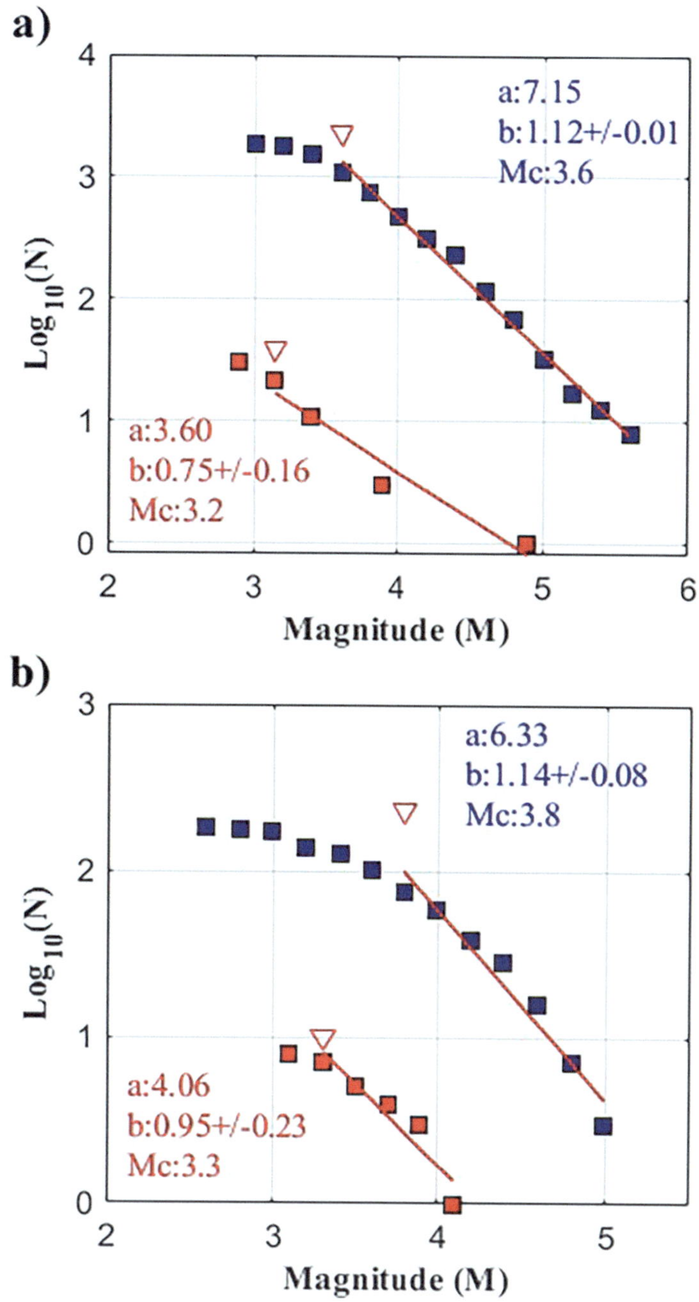

Fig. 4.9 b-value computed from foreshocks (brown) and aftershocks (blue) of Sichuan **a** 2008-05-12 earthquake, **b** Sichuan 2013-04-20 earthquake. The inverted triangles denote the magnitude completeness (Mc)

References

Anderson RE (1971) Thin skin distension in Tertiary rocks of southeastern Nevada. Bull Geol Soc Am 82:43–58

Archer C, Allen JR (1969) A catalogue of earthquakes in the Lake Kariba area, 1959–1968. Meteorological Services, Salisbury, 35 pp

Beal T (2014) When Lake Mead rapidly filled up, quakes followed. Arizona Daily Star, 25 Dec 2011. Updated 23 July 2014

Bell ML, Nur A (1978) Strength changes due to reservoir-induced pore pressure and stresses and application to Lake Oroville. J Geophys Res 83:4469–4483

Bond G (1953) The origin of thermal and mineral waters in the middle Zambezi Valley and adjoining territory. Trans Geol Soc S Afr 56:131–148

Bond G (1960) The geology of the middle Zambezi Valley. Unpublished manuscript

Carder DS (1945) Seismic investigations in the Boulder Dam area, 1940–1944, and the influence of reservoir loading on earthquake activity. Bull Seismol Soc Am 35:175–192

Chen L, Talwani (1999) Reservoir-induced seismicity in China. In: Talebi S (ed) Seismicity caused by mines, fluid injections, reservoirs, and oil extraction. Pageoph topical volumes. Birkhäuser, Basel, pp 133–149. https://doi.org/10.1007/978-3-0348-8804-2_8

Chen H, Xu Z, Li M (2010) The relationship between large reservoirs and seismicity. Int Water Power Dam Constr 29–33. www.waterpowermagazine.com

Chen WP, Hsu L (2013) Historic Seismicity near the source zone of the great 2008 Wenchuan earthquake: Implications for seismic hazards. Tectonophysics, 584:114–118.

Committee of Experts (1968) Report on the Koyna Earthquake of December 11, 1967—1 and 2. Government of India Press, New Delhi, pp 75
Gahalaut K, Gahalaut VK (2010) Effect of the Zipingpu reservoir impoundment on the occurrence of the 2008 Wenchuan earthquake and local seismicity. Geophys J Int 183(1):277–285
Galanopoulos AG (1967) The large conjugate fault system and the associated earthquake activity in Greece. Ann Geol Pays Helleniques (Athens) 18:119–134
Geological Survey of India (1968) A geological report on the Koyna earthquake of 11 December 1967. Calcutta, 242 pp
Guha SK, Gosavi PD, Varma MM, Agarwal BNP, Padale JG, Marwadi SC (1970) Recent seismic disturbances in the Shivajisagar Lake area of Koyna hydroelectric project, Maharashtra, India, 2. Rep. C. W. P. R. S. 25 pp
Gupta HK (1983) Induced seismicity hazard mitigation through water level manipulation at Koyna, India: a suggestion. Bull Seism Soc Am 73:679–682
Gupta HK (1992) Reservoir Induced Earthquakes. Elsevier Scientific Publishing Company, Amsterdam, p 364
Gupta HK (2017) Koyna, India, an ideal site for near field earthquake observations. J Geol Soc India 90(6):645–652
Gupta HK (2018) Review: reservoir triggered seismicity (RTS) at Koyna, India over the past 50 years. Bull Seism Soc Am 108(5B):2907–2918
Gupta HK (2022) Artificial water reservoir-triggered seismicity (RTS): most prominent anthropogenic seismicity. Surv Geophys. https://doi.org/10.1007/s10712-021-09675-z
Gupta HK, Rajendran K (1986) Large artificial water reservoirs in the vicinity of the Himalayan foothills and reservoir induced seismicity. Bull Seism Soc Am 76(1):205–215
Gupta HK, Rekapalli R (2022) Mw $\geq$ 5 aftershocks of the 2008 Sichuan earthquake: analysis of temporal variation of Omori Law p-value. Front Earth Sci. https://doi.org/10.3389/feart.2022.964245
Gupta HK, Rekapalli R (2026) The 2008 Mw 7.9 Wenchuan, China earthquake: not a case of reservoir triggered seismicity. Gondwana Research 151:184–188
Gupta HK, Rastogi BK, Narain H (1972a) Common features of reservoir associated seismic activities. Bull Seismol Soc Am 62:481–492
Gupta HK, Rastogi BK, Narain H (1972b) Some discriminatory characteristics of earthquakes near Kariba, Kremasta, Koyna artificial lakes. Bull Seismol Soc Am 63:493–507
Gupta HK, Purnachandra Rao N, Shashidhar D, Mallika K (2008) The disastrous M 7.9 Sichuan earthquake of 12 May 2008. J Geol Soc India 72:325–330
Hatzfeld D, Nord J, Paul A, Guiguet R, Broille P, Rugg JC (1995) The Kozani-Grevena (Greece) earthquake of May 13, 1995, Ms = 6.6: preliminary results of a field multidisciplinary survey. Seismol Res Lett 66(6):61–70
Hitchon B (1958) The geology of the Kariba area. North Rhod Geol Surv Rep 3:41
Hsu T-H, Lo H-H, Ho T-H, Yang C-J (1976) Strong-motion observation of water induced earthquakes at Hsinfengkiang reservoir in China. Eng Geol 10(2–4):315–330
Ishikawa Y, Oike K (1982) On reservoir-induced earthquakes in China. Zishin 35(2):171–181
Jones AE (1944) Earthquake magnitudes, efficiency of stations and perceptibility of local earthquakes in the Lake Meade area. Bull Seismol Soc Am 34:161–173
Knill JL, Jones KS (1965) The recording and interpretation of geological conditions in the foundations of the Rosieres, Kariba and Latiyan Dams. Geotechnique 158:94–124
Krishnan MS (1960) Geology of India and Burma. Higginbotham, Madras, 604 pp
Lin P, Wong RK, Li QB, Zhou WY (2009) Effect analysis of structural safety of typical large dams from Wenchuan 8.0 earthquake. Chin J Rock Mech Eng 28(6):1261–1269
Longwell CR (1936) Geology of the Boulder reservoir floor, Arizona, Nevada. Bull Geol Soc Am 47:1393–1476
Longwell CR (1963) Reconnaissance geology between Lake Mead and Arvis Dam, Arizona, Nevada. U.S. Geol Surv Prof Pap 374-E:51
Michas G, Pavlou K, Vallianatos F, Drakatos G (2020) Correlation between seismicity and water level fluctuations in the Polyphyto Dam, North Greece. Pure Appl Geophys. https://doi.org/10.1007/s00024-020-02465-50
Mogi K (1963) Some discussions on aftershocks, foreshocks and earthquake swarms—the fracture of a semi-infinite body caused by an inner stress origin and its relation to earthquake phenomena. Bull Earthq Res Inst 41:615–658
Narain H, Gupta HK (1968) Koyna earthquake. Nature 217:1138–1139
Papazachos BC, Papazachou CB (1989) The earthquakes of Greece. Ziti Publications, Thessaloniki, p 304
Pavlou K (2019) Relationship between observer seismicity and water level fluctuations in Polyphyto Dam area (North Greece). J Geogr Environ Earth Sci Int 21(2):1–10
Pavlou K (personal communication, 2021)
Pavlou K, Kaviris G, Chousianitis K, Drakatos G, Kouskouna V, Makropoulos K (2013) Seismic hazard assessment in Polyphyto Dam area (NW Greece) and its relation with the “unexpected” earthquake of 13 May 1995 (Ms = 6.5, NW Greece). Nat Hazards 13:141–149
Pavlou K, Drakatos G, Kouskouna V, Makropoulos K, Kranis H (2016) Seismicity study in Pournari reservoir area (W. Greece). J Seismol 20:701–710
Raphael JM (1954) Crustal disturbance in the Lake Mead area. In: Engineering monographs no. 14. U.S. Bureau of Reclamation, Denver, CO, p 14
Rogers AM, Gallanthine SK (1974) Seismic study of earthquakes in the Lake Mead region. Environmental Research Corporation, U.S.A., Rep. Contract 14-08-0001-13069
Shen C, Chang C, Chen H, Li T, Hueng L, Wang T, Yang C, Lo H (1974) Earthquakes induced by reservoir impounding and their effect on the Hsinfengkiang Dam. Sci Sin 17(2):239
Simpson DW (1986) Triggered earthquakes. Annu Rev Earth Planet Sci 14:21–42
Snow DT (1972) Geodynamics of seismic reservoirs. In: Proceedings of symposium on percolation through fissured rock. Ges. Erd-und Grundbau, Stuttgart, T2J, pp 1–19
Snow DT (1974) The geologic hydrologic and geomorphic setting of earthquakes at Lake Kariba. Paper presented at the international colloquium on seismic effects of reservoir impounding, Mar 1983. Royal Society, London
Stein S, Wiens DA, Fujita K (1982) The 1966 Kremasta reservoir earthquake sequence. Earth Planet Sci Lett 59:49–60
Tao W, Masterlark T, Shen ZK, Ronchin E (2015) Impoundment of the Zipingpu reservoir and triggering of the 2008 Mw 7.9 Wenchuan earthquake, China. J Geophys Res Solid Earth 120:7033–7047. https://doi.org/10.1002/2014JB011766
Wang M, Yang M, Hu Y, Li T, Chen Y, Chin Y, Feng J (1976) Mechanism of the reservoir impounded earthquakes at Hsinfengkiang and a preliminary endeavour to discuss their cause. Eng Geol 10:331–351
Wieland M (2010) Seismic aspects of large dams: lessons learnt from the May 12, 2008 Wenchuan earthquake. In: Proceedings of the international symposium on dams & sustainable water resources development, 78th ICOLD annual meeting, Hanoi, Vietnam, 20–26 May 2010
Wieland M (2017) Reservoir-triggered seismicity and effect on seismic design criteria for large storage dam projects. In: 16th world conference on earthquake, 16WCEE 2017, Santiago Chile, pp 1–12
Zhang C (2011) The performance of dams during the Wenchuan 5–12 earthquake and lessons learned from the event. J Earthq Tsunami 5(4):309–327
Zhu C, Liu Y, Wang C, Lu R, Chen J (1982) Modeling focal parameters for the magnitude 5.3 earthquake of the Xinfengjiang Reservoir area, People's Republic of China. Bull Seismol Soc Am 72(4):1085–1092

5 RTS Mechanism

Evans (1966) study of the injection-induced earthquakes, that occurred in the vicinity of the Rocky Mountain Arsenal near Denver, Colorado, USA, was the first one which showed that such anthropogenic activities could lead to triggering of the earthquakes. Evans (1966) had made use of the Hubbert and Rubey's (1959) inference that the increased pore fluid pressure due to fluid injection was responsible for inducing the earthquakes. One of the first to point out the role of reservoir loading in triggering earthquakes were Gough and Gough (1970a, b) while studying the triggered earthquakes at Lake Kariba on Zambia-Zimbabwe border. Later Bell and Nur (1978) and Roeloffs (1988) addressed the role of artificial water reservoirs in triggering earthquakes. It was pointed out that water load induced stress at focal depth of earthquakes is very small and it could only provide a trigger to an already critically stressed strata causing earthquakes. Gupta et al. (1972a, b) undertook study of the available cases of artificial water reservoir triggered earthquakes and identified factors such as the rate of increase in water levels, the highest water levels reached and the duration of retention of the high water levels as the factors influencing RTS. The role of fluid pressure in triggering earthquakes was investigated by Snow (1972). However, he made use of very simplified model of a reservoir with finite depth but infinite width. This was the beginning which was later followed by considering more realistic models by Withers and Nyland (1976) as well as by Bell and Nur (1978). Biot's (1941) consolidation theory was used in these models (Rice and Cleary 1976).

The role of lake water level changes in triggering the earthquakes was addressed by Simpson and Negmatullaev (1981) at Nurek Dam in Tadjik Republic. Zoback and Hickman (1982) conducted well planned experiments dealing with in-situ stress measurements of physical properties of the rock strata that controlled the triggered earthquakes at Montecillo, South Carolina, USA.

Detailed investigations of the rate of loading at Koyna, India carried out by Gupta (1983) showed that all $M \geq 5$ events occurred when the rate of loading exceeded 40 ft/week. It was found that a rate of loading of 40 ft/week was a necessary but not a sufficient condition for $M \geq 5$ earthquakes to occur. It was demonstrated that through appropriate monitoring of the rate of loading at Koyna, India could possibly avoid the occurrence of $M \geq 5$ earthquakes.

Talwani and Acree (1984/85) examined the role of pore pressure diffusion in triggering of the earthquakes. They conducted experiments to measure hydraulic diffusivity and Skempton's coefficient at the Bad Creek Reservoir in South Carolina, USA (Talwani et al. 1999). The water levels in an observation well located 250 m away from the reservoir had direct correspondence with the water levels in the reservoir. To start with, there was a delay of 96 h between the reservoir level and the corresponding level in the observation well. Later this delay reduced to 72 h and was stabilized. Talwani et al. (2007) conducted frequency independent estimate of diffusivity, and it was found to be ~ $0.076\ m^2\ s^{-1}$. At the same time, a Skempton coefficient of 0.66 was found. Both these estimates were for undrained response of the reservoir. Talwani et al. (2007) found diffusivity varying between 0.1 and $10\ m^2\ s^{-1}$ while examining some 90 case histories: this range of diffusivity corresponds to intrinsic permeability of 5×10^{-16} and 5×10^{-14}. Fault instability changes and their correspondence to repeated loading and unloading of the reservoir was addressed by Roeloffs (1988). The role of the rock property inhomogeneities in triggering the earthquakes was addressed by Simpson and Narsimha (1990).

This chapter deals with the current knowledge of RTS mechanism and is based on Gupta (2022): https://doi.org/10.1007/s10712-021-09675-z.

H. K. Gupta, *Artificial Water Reservoir Triggered Seismicity*, https://doi.org/10.1007/978-3-032-05873-7_5

Bell and Nur (1978) addressed the three main consequences of loading an artificial water reservoir. These are:

1. Following the filling of a reservoir, the elastic stress increases.
2. Due to increase in elastic stress, fluid pressure in saturated rocks increases, as the pore volume decreases due to compaction.
3. In response to fluid migration, the pore pressure changes. At times the local water table is low before the impoundment of a reservoir. Following the impoundment, the water table rises, as the water flows from the reservoir to the unsaturated strata.

The deviatory effect of pore fluid pressure changes in anisotropic rocks was investigated by Chen and Nur (1992). Previous work mostly addressed isotropic rocks. This has been a very useful development as most of the reservoirs would have anisotropic rocks. Chander and Kalpna (1997) addressed fault instability in response to interplay of mobilized frictional stresses and the shear stresses acting on a fault. Later Kalpna and Chander (2000) developed an algorithm for simulating pore pressure and stresses for realistic 3D models of reservoirs.

A diffusion model has been presented by Pandey and Chadha (2003) for the Koyna region, assuming a highly permeable near-vertical fault zone in the basement below the Deccan traps, connected to the reservoir through fissures, fractures and vesicles in the overlying basalts. Propagation and distribution of excess head, i.e., fraction of surface loading in an inhomogeneous medium, were examined with the assumption of a conductive vertical fault having hydraulic conductivity of about 2.0 m per day. The vertical fault with chosen permeability facilitates the diffusion process in such a way that excess pressure front equivalent to 5–15% of the initial head on the surface can reach up to the hypocentral depth of 6–8 km. They infer that a small-stress perturbations of 0.75–2.25 bar at hypocentral depths is sufficient to cause failure on critically stressed preexisting faults in the Koyna region. The continued RTS at Koyna is comprehended by Gahalaut et al. (2004) due to fault interaction through stress transfer. They opined that strike–slip faults of the Koyna region and normal faults in Warna region can be brought closer to failure with change in Coulomb Failure Stress (ΔCFS) between 5 and 10 kPa. A homogeneous poroelastic model for the period before the M 6.3 earthquake on 10 December 1967 earthquake has been presented by Gavrilenko et al. (2010). They calculated stress field resulting from the water-level fluctuation and the pore pressure changes due to the undrained and diffusive response of the medium assuming the diffusivity 0.2 m^2/s and a Skempton coefficient of 0.31. The average Coulomb stress threshold for triggering earthquakes is around 0.04 MPa according to this model. Compared to pre-M 6.3 event, a tenfold increase in hydraulic conductivity (2.5 m^2/s) is required to match the aftershock decay rate. However, this model could not explain the events that occurred some 9 months after the main shock. Consequently, a hydromechanical model with two compartments having contrasting diffusivities was inferred, and the main shock was inferred to cause enhancement in fluid flow south of the Koyna Dam. In a 3D modeling for pore pressure diffusion at Koyna, Yadav et al. (2017) found stress perturbation of 80–147 kPa sufficient to trigger earthquakes in the region. Hazarika et al. (2017) performed 2D modeling to estimate the role of permeability in modeling of RTS in Koyna region. Based on poroelastic theory, they used a 2D finite element model to simulate pore pressure development after 5 years of impoundment of Koyna Dam in 1962 using a range of permeability of 10^{-16}–10^{-14} based on the information of permeability of basalt from the Koyna region. This permeability does not produce an appreciable change at 2 km depth. A tenfold increase in permeability is required to cause critical pore pressure changes of 0.1–1.0 MPa up to a depth of 10 km, whereas study of core samples to a depth of 1522 m in the Koyna region provided evidence of fractured zones, providing the channels for water flow to the focal depths and thereby causing triggering.

Do-Nascimento et al. (2005) inferred the observed RTS at Açu Reservoir located in north-east Brazil through diffusion of pore pressure. The 3D numerical modelling was used to explain diffusion of the pore pressure. It was found that heterogeneous hydraulic properties of the prevailing fault system around the Açu reservoir could be inverted using the distribution of RTS events in space and time.

Cheng et al. (2016) have used a fully coupled 3D poroelastic finite element model to explain RTS at Aswan Dam in Egypt. They found that the pore pressure and Coulomb stress on earthquake hosting fault zone increase with the passage of time. The impoundment of the Aswan Reservoir appears to have been responsible for triggering M 5.7 earthquake: this inference has been based on the magnitude of Coulomb failure stress change (ΔCFS) estimated to be 0.1 MPa at the hypocenter of the 1981 M 5.5 RTS event. It was estimated that the contribution to the total ΔCFS, by the elastic load is less than 3%. A maximum accumulated surface deformation of 80 cm beneath the Aswan reservoir is estimated from the time of its impoundment till the occurrence of the 14 November 1981 M 5.7 earthquake. It is estimated that the impoundment of the Aswan Lake may have caused a total accumulation of 1.0×10^{10} J strain energy. This is just a few percentages of the total energy released at Aswan due to RTS, including the M 5.7 14 November 1981 main RTS event. This is in support of the argument that filling of an artificial water reservoir only provides a trigger for RTS.

5.1 Triggered or Induced Earthquakes

For a long time the adjectives, 'triggered' and 'induced' have been used interchangeably. It is important and necessary to distinguish between the two. A solution was provided by McGarr and Simpson (1997). According to them the adjective 'triggered' should be used when the causative factor, such as the building of an artificial water reservoir, is responsible for a small percentage of the total energy released in the consequent occurrence of earthquakes. It was pointed out that creation of the deepest artificial water reservoirs causes a stress change of the order of 1 MPa. Therefore, it is appropriate to classify earthquakes occurring consequent to filling of the artificial water reservoirs as '*triggered*'.

McGarr and Simpson (1997) raised four essential questions that should be addressed to determine whether an event is '*triggered*' or '*induced*'. These are:

1. What is the essential mechanism causing the seismicity?
2. Is the corresponding stress change large or small compared to typical seismic stress drop (e.g., Brune 1970)?
3. Does the simulated seismicity fall in the category of 'triggered' or 'induced'? and,
4. What factors determine the maximum size of the stimulated earthquake?

Following observations were made by McGarr and Simpson (1997):

1. The causal mechanism for RTS can involve all the three factors, namely (a) purely increase in shear and normal stress, (b) pore pressure increase due to diffusion, and (c) coupled poroelastic increase in the pore pressure due to compaction.
2. Even the deepest reservoirs impose a load of only a few MPa, when extrapolated to depths of seismicity, it is of the order of 0.1 MPa, which is significantly smaller, compared to the stress drop during the triggered earthquakes being of the order of 1.0–10 MPa.
3. The low induced stresses place reservoir seismicity in the category of '*triggered*' seismicity.
4. The maximum size of the triggered earthquake would depend on the prevailing stress situation in the reservoir region. For M ~ 5 earthquakes, the preexistence of adequately large faults in the vicinity of the reservoir is requisite.

Foulger et al. (2018) has questioned the classification made by McGarr and Simpson (1997). They argue that all earthquakes release some pre-stored strain energy and hence all anthropogenic earthquakes must be classified as '*induced*'.

5.2 Rapid, Delayed and Continued Seismicity

Simpson et al. (1988) grouped RTS in two broad categories, namely 'rapid response' and 'delayed response'. The 'rapid response' category is the one where RTS is observed immediately after the impoundment of the reservoir. This has been observed globally at RTS sites like Nurek, Tajikistan; Kariba, Zambia-Zimbabwe border; Kremasta, Greece and at several other reservoir sites. At these sites RTS occurred soon after the impoundment of the reservoir or following rapid changes in water levels. Under the category of 'delayed response' main/major events occur after a few cycles of loading and unloading of the reservoir. Koyna, India; Aswan, Egypt fall in this category. It was also observed by Simpson et al. (1988) that there are cases where both, the 'rapid response' and the 'delayed response' are observed. In the 'delayed response' peak RTS is observed with short-term water level changes when the water level is at the peak. 'Delayed response' RTS is related with the pore pressure diffusion associated with the loading of the reservoir and often located away from the reservoir. Particularly, in the case of the Aswan Reservoir, Egypt, the main earthquake occurred 17 years after the impoundment of Lake Nasser at a depth of ~ 20 km. Such focal depths are not usually associated with RTS. It is noticeable that the 1981 main earthquake occurred when the water level in the reservoir was at the peak (Gupta 1992).

Two temporal categories were introduced by Talwani (1997). These are 'initial seismicity' and 'protracted seismicity'. The 'initial seismicity' includes cases where RTS is observed soon after the impoundment of the artificial water reservoir, or when the water level maxima in the reservoir exceeds the previous maxima. The 'protracted seismicity' continues for many years.

Talwani carried out two-dimensional calculations as were reported by Roeloffs (1988) he infers that 'protracted seismicity' is influenced by several factors including lake level fluctuations, reservoir dimensions, hydromechanical properties of media etc.

In a detailed work, Gupta (2002) classified RTS into three categories, namely, 'rapid response', 'delayed response' and 'continued seismicity'. He pointed out that there are RTS sites which are characterized one of the three or a combination of two or three categories. 'Continued seismicity' was observed by Gupta and Combs (1976) for the Koyna, India. As of now, it appears that Koyna is a typical example of 'rapid', 'delayed' and 'continued' seismicity. RTS started soon after the impoundment of Shivaji Sagar Lake in 1962 qualifying it to be a case of 'rapid response'. The largest M 6.3 earthquake occurred 5 years later in 1967, amounting to 'delayed response' and RTS has continued till now (2025) qualifying it to be the case of 'continued seismicity'. It may, however, be mentioned, that the level of seismicity has reduced by several folds compared to 1967 (Gupta 2018).

5.3 How Long Will RTS Continue?

Most of the RTS sites, globally, are located in the stable continental regions (SCRs). There is a huge difference in the rate of strain accumulation in SCRs, intraplate regions and plate boundaries, being 10^{-9}–10^{-10}, 10^{-7}–10^{-9}, and 10^{-5} respectively (Gupta and Johnston 1998).

It is generally observed that RTS gets initiated soon after the impoundment of a reservoir. Sometimes it starts after a few cycles of loading and unloading of the reservoir. Mostly the largest RTS event has occurred within a few years of the initial loading of the reservoir. Here we are examining 4 cases of RTS where the largest RTS event of M $\geq$ 6 has occurred. The cases discussed here are Hsinfengkiang, China; Kariba, Zambia-Zimbabwe border; Kremasta, Greece; and Koyna, India. The case of Polyphyto Dam, Greece is not included here since after building the Polyphyto Dam in 1974, other dams in the near vicinity were built in the years to follow. The latest dam was added in 2019 (Michas et al. 2020).

The Hsinfengkiang Dam in China was built in 1959. The largest RTS event of M 6.1 occurred in 1961. At Hsinfengkiang, soon after impounding, thousands of events occurred. The number of events occurring in the vicinity of the Hsinfengkiang reservoir continued to decrease. By 1978, there were just a few events occurring every month, as reported by Ishikawa and Oike (1982) and Gupta (1992). Filling of Lake Kariba on Zambia-Zimbabwe border took place in 1958. This was followed by a large number of earthquakes occurring leading to the occurrence of M 6.2 on the 13 April 1963. However, by 1974, the seismic activity in the vicinity of Lake Kariba dropped to just 10% of 1963 level (Gough and Gough 1970a, b). The Lake Kremasta, Greece has the distinction of being one of the 5 sites globally where an RTS event of M $\geq$ 6 has occurred. It was impounded in 1965. The filling was very rapid and peak levels were reached in 1966. There was immense seismic activity and M 6.2 earthquake occurred on 5 February 1966 (Galanopoulos 1967; Stein et al. 1982). The RTS at Kremasta was very short lived and was over in about 6 months' time. As far as Koyna, India is concerned, RTS started soon after the impoundment of the Shivaji Sagar Lake in 1962. The first M $\geq$ 5 earthquake occurred on 13 September 1967, soon to be followed by the damaging M 6.3 earthquake on the 10 December 1967. The activity has decreased considerably by now. The last M ~ 4 earthquake occurred on 20 June 2019.

The details of RTS at these 4 sites are provided in Table 5.1.

Gupta et al. (2002) addressed the question as to how long the RTS will continue at Koyna. Based on the global SCR earthquake data and low strain rates reported by Johnston et al. (1994) it was estimated that an M 6.8 event is possible for the Koyna region. The Koyna region was stressed close to critical before the impoundment of the reservoir. It was inferred that the heterogeneity in the region has increased from the study of 'b' values, foreshock-aftershock patterns, and the ratio of the largest aftershock to the main shock magnitude. When the study was conducted by Gupta et al. (2002), 17 M $\geq$ 5 earthquakes and several hundred smaller earthquakes had occurred. It was estimated that about one-half energy of a M 6.8 earthquakes had been already released. It was inferred that triggered earthquakes would continue for another 3–4 decades. However, no M ~ 6 earthquake would occur, and the occurrence of M ~ 5 earthquakes would be governed by Kaiser (1953) effect. With the passage of time, the RTS at Koyna, India has decreased. Till the end of the year 2000,

Table 5.1 RTS duration at 4 sites where M $\geq$ 6 earthquakes occurred

Reservoir	Impoundment	Largest RTS event (year)	How long RTS continued (10% of the maximum)
Hsinfengkiang (China)	1959	6.1 (1962)	Late 1970s
Kariba (Zambia-Zimbabwe)	1958	6.2 (1963)	1974
Kremasta (Greece)	1965	6.2 (1966)	1967
Koyna (India)	1962	6.3 (1967)	2021

there were 17 M $\geq$ 5 earthquakes. However, by the end of 2024 only 4 additional M $\geq$ 5 earthquakes occurred. The last M $\geq$ 5 earthquake occurred on the 12 December 2009.

Anuradha et al. (2021) provided a very good proof of the Koyna region being stressed close to critical when they discovered that hammer drilling conducted for the Pilot Borehole had triggered earthquakes in the immediate vicinity of the site.

References

Anuradha MU, Rekapalli R, Sudheer K, Jagadeesh P, Satyanarayana HVS (2021) Triggering of earthquakes at Koyna, India and hammer drilling for the pilot borehole. Nat Hazards 1–9

Bell ML, Nur A (1978) Strength changes due to reservoir-induced pore pressure and stresses and application to Lake Oroville. J Geophys Res 83:4469–4483

Biot MA (1941) General theory of three-dimensional consolidation. J Appl Phys 12:155–164

Brune JN (1970) Tectonic Stress and Spectra of seismic shear waves from earthquakes. J Geophys Res 75: 4997–5009

Chander R, Kalpna (1997) On categorizing induced and natural tectonic earthquakes near new reservoirs. Eng Geol 46

Chen Q, Nur A (1992) Pore fluid pressure effects in an isotropic rocks: mechanisms of induced seismicity and weak faults. Pure Appl Geophys 139:463–480

Cheng H, Zhang H, Shi Y (2016) High resolution numerical analysis of the triggering mechanism of M_L 5.7 Aswan Reservoir earthquake through fully-coupled poroelastic finite element modeling. Pure Appl Geophys 173(5):1593–1605

Do-Nascimento AF, Lunn RJ, Cowie PA (2005) Numerical modeling of pore pressure diffusion in a reservoir-induced seismicity site in northeast Brazil. Geophys J Int 160:249–262

Evans MD (1966) Man-made earthquakes in Denver. Geotimes 10:11–17

Foulger GR, Wilson MP, Gluyas JG, Julian BR, Davies RJ (2018) Global review of human-induced earthquakes. Earth Sci Rev 178:438–514

Gahalaut VK, Gahalaut K, Singh SK (2004) Fault interaction and earthquake triggering in the Koyna-Warna region, India. Geophys Res Lett. https://doi.org/10.1029/2004GL019818

Galanopoulos AG (1967) The large conjugate fault system and the associated earthquake activity in Greece. Ann Geol Pays Helleniques 18:119–134

Gavrilenko P, Singh C, Chadha RK (2010) Modeling the hydromechanical response in the vicinity of the Koyna reservoir (India): results for the initial filling period. Geophys J Int 183(1):461–477

Gough DI, Gough WI (1970a) Stress and deflection in the lithosphere near Lake Kariba, 1. Geophys J 21:65–78

Gough DK, Gough WI (1970b) Load induced earthquakes at Kariba, 2. Geophys J R Astron Soc 21:79–101

Gupta HK (1983) Induced seismicity hazard mitigation through water level manipulation: a suggestion. Bull Seismol Soc Am 73:679–682

Gupta HK (1992) Reservoir Induced Earthquakes, Elsevier Scientific Publishing Company, Amsterdam p 364

Gupta HK (2002) A review of recent studies of triggered earthquakes by artificial water reservoirs with special emphasis on earthquakes in Koyna, India. Earth Sci Rev 58:279–310

Gupta HK (2018) Review: reservoir triggered seismicity (RTS) at Koyna, India over the past 50 years. Bull Seismol Soc Am 108(5B):2907–2918

Gupta HK (2022) Artificial water reservoir-triggered seismicity (RTS): most prominent anthropogenic seismicity. Surv Geophys 43:619–659. https://doi.org/10.1007/s10712-021-09675-z

Gupta HK, Combs J (1976) Continued seismic activity at the Koyna reservoir site. India Eng Geol 10:307–313

Gupta HK, Johnston AC (1998) Note on Chapman conference on stable continental region (SCR) earthquakes. J Geol Soc India 52:115–117

Gupta HK, Rastogi BK, Narain H (1972a) Common features of reservoir associated seismic activities. Bull Seismol Soc Am 62:481–492

Gupta HK, Rastogi BK, Narain H (1972b) Some discriminatory characteristics of earthquakes near Kariba, Kremasta, Koyna artificial lakes. Bull Seismol Soc Am 63:493–507

Gupta HK, Mandal P, Rastogi BK (2002) How long will the triggered earthquakes at Koyna, India continue? Curr Sci 82:202–210

Hazarika P, Yadav A, Roy S (2017) Influence of permeability in modeling of reservoir triggered seismicity in Koyna region, western India. J Geol Soc India 90(6):728–732

Hubbert MK, Rubey WW (1959) Role of fluid pressure in mechanics of over thrust faulting, 1. Bull Geol Soc Am 70:115–166

Ishikawa M, Oike K (1982) On reservoir-induced earthquakes in China. Zishin 35(2):171–218

Johnston AC, Kanter LR, Coppersmith KJ, Cornell CA (1994) The earthquakes of stable continental regions. Volume 3: seismicity data sheets (part 1); final report no. EPRI-TR-102261-V3. Electric Power Research Institute

Kaiser J (1953) Erkenntnisse and Folgerungen aus der Messung von Gerauschen bei Zugbeanspruchung von metallischen Werkstoffen. Arch Eisenhuttenwes 24:43–45 (in German)

Kalpna, Chander R (2000) Green's function-based stress diffusion solutions in the porous elastic half space for time varying finite reservoir loads. Phys Earth Planet Inter 120:93–101

McGarr A, Simpson D (1997) Keynote lecture: a broad look at induced and triggered seismicity "Rock bursts and seismicity in mines." In: Gibowicz SJ, Lasocki S (eds) Proceedings of 4th international symposium on rock bursts and seismicity in mines, Poland, 11–14 Aug 1997. A. A. Balkema, Rotterdam, pp 385–396

Michas G, Pavlou K, Vallianatos F, Drakatos G (2020) Correlation between seismicity and water level fluctuations in the Polyphyto Dam, North Greece. Pure Appl Geophys. https://doi.org/10.1007/s00024-020-02465-50

Pandey AP, Chadha RK (2003) Surface loading and triggering earthquakes in the Koyna-Warna region, western India. Phys Earth Planet Inter 139:207–223

Rice JR, Cleary MP (1976) Some basic stress diffusion solutions for fluid-saturated elastic porous media with compressible constituents. Am Geophys Union Rev Geophys Space Phys 14(2):227–242

Roeloffs EA (1988) Fault stability changes induced beneath a reservoir with cyclic variations in water level. J Geophys Res 93(B3):2107–2124

Simpson DW, Negmatullaev SK (1981) Induced seismicity at Nurek Reservoir, Tadjikistan, USSR. Bull Seismol Soc Am 71(5):1561–1586

Simpson DW, Leith WS, Scholz CH (1988) Two types of reservoir-induced seismicity. Bull Seismol Soc Am 78(6):2025–2040

Simpson DW, Narsimhan (1990) Inhomogenetics in rock properties and their influence on reservoir induced seismicity. Gerlands Beitr Geophys 99:205–219

Snow DT (1972) Geodynamics of seismic reservoirs. In: Proceedings of symposium on percolation through fissured rock. Ges. Erd-und Grundbau, Stuttgart, T2J, pp 1–19

Stein S, Wiens DA, Fujita K (1982) The 1966 Kremasta reservoir earthquake sequence. Earth Planet Sci Lett 59:49–60

Talwani P (1997) On the nature of reservoir-induced seismicity. Pure Appl Geophys 150:473–492

Talwani E, Acree S (1984/85) Fore pressure diffusion and the mechanism of reservoir induced seismicity. Pageoph 122:947–965

Talwani P, Cobb JS, Schaeffer MF (1999) In situ measurements of hydraulic properties of a shear zone in northwestern South Carolina. J Geophys Res 104(B7):14993–15003

Talwani P, Chen L, Gahalaut K (2007) Seismogenic permeability. J Geophys Res 112:B07309, 1–18. https://doi.org/10.1029/2006JB004665

Withers RJ, Nyland E (1976) Theory for the rapid solution of ground subsidence near reservoirs on layered and porous media. Eng Geol 10:169–185

Yadav A, Gahalaut K, Rao NP (2017) 3-D modeling of pore pressure diffusion beneath Koyna and Warna reservoirs, western India. Pure Appl Geophys 174(5):2121–2132

Zoback MD, Hickman S (1982) Physical mechanisms controlling induced seismicity at Monticello Reservoir, South Carolina. J Geophys Res 87:6959–6974

6 Koyna, India (1962–2004)

Among the various cases of RTS, globally, Koyna, India is one of the most significant sites where an M 6.3 RTS event occurred on 10 December 1967 (Gupta, 1992, 2002). It claimed about 200 human lives. The nearby Koyna Township was in shambles, where almost 80% houses were destroyed. The city of Bombay located some 230 km from Koyna Dam was rocked. People ran outside their homes for safety. The electric power was off due to Koyna Hydroelectric Plant being out of operation.

The Shivaji Sagar Lake was created by the construction of the Koyna Dam in 1962 which is located in the Peninsular Shield of India. The seismic zoning map for India prepared in 1962 and revised in 1966 characterize the region to be aseismic. There is no reported earthquake in the vicinity of the Koyna before the impoundment of the Shivaji Sagar Lake by the Koyna Dam. The Pune seismic station located about 110 km from the Koyna Dam, started functioning in 1935. With the installed Milne Shaw seismograph there, any earthquake of M $\geq$ 3 occurring in the Koyna region would have been recorded at Pune. No such earthquake was recorded at Pune. Soon after the impoundment of the Shivaji Sagar Lake, tremors in the vicinity of the Koyna Dam became prevalent, often accompanied with sound. The strongest of these tremors would rattle the windows of the houses in Koyna Nagar (Mane 1967). A network of 4 observatories was set up to monitor seismic activity in the vicinity of the Koyna Dam. The hypocenters of these tremors were found to be clustering in the vicinity of the Shivaji Sagar Lake. These were small events. By the end of 1969, the network of the seismic stations was increased to 7 stations (Guha et al. 1974a, b). Five earthquakes that preceded the 10 December 1967 M 6.3 earthquake were strong enough to be recorded by the entire Indian Seismic Station Network. The strongest of these was the 13 September 1967 M 5.7 earthquake. The 10 December 1967 earthquake of M 6.3 occurred at 22.51 Universal Time (04:51 Indian Standard Time).

6.1 Focal Parameters of the 10 December 1967 Koyna Earthquake

The focal parameters of the 10 December 1967 Koyna earthquake were estimated by a several agencies. Table 6.1 provides a gist of the same. There are major variations in the estimates of these parameters. Gupta et al. (1971) observed that the various origin time and the epicenters obtained by different agencies fit well with a multiple seismic event. They conducted a detailed analysis of seismograms obtained from 48 seismic stations globally as well as the correlograms from the seismic arrays located at Warramunga (WRA), Australia; Yellowknife (YKA), Canada; and Eksdalemuir (EKA), United Kingdom. Thirlaway (1963) had pointed out the differences in the correlograms obtained from seismic arrays for earthquakes, explosions and complex earthquake sources. The correlograms for simple earthquakes and explosions are characterized by a large built up of the energy which gradually dies, whereas the complex earthquakes correlograms have bursts of energy at short interval of time corresponding to the release of energy by progressive rupture of a complex earthquake source. The Koyna earthquake correlograms (Fig. 6.1) clearly fall in the category of a complex source earthquake. A comparison of the correlograms of an explosion, a simple earthquake and a complex earthquake (after Thirlaway 1963) is provided in Fig. 6.2. Among the six events marked on these correlograms (Fig. 6.3), the second and the third event occurred 2 and 5 s after the first event which are most conspicuous. These could be very well identified on the correlograms as well as the short (Fig. 6.4) and long period (Fig. 6.5) seismograms of 48 WWSSN stations globally. The initiation of the Koyna multiple seismic events is taken as the epicenter determined by the network of C.W.P.R.S. Seismic Stations (Table 6.1), and the approximate distances of the 2nd and the 3rd event has been calculated using the following

H. K. Gupta, *Artificial Water Reservoir Triggered Seismicity*, https://doi.org/10.1007/978-3-032-05873-7_6

relation given by Wyss and Brune (1967), the seismic stations being arranged azimuth wise.

$$\Delta i = \frac{t_{\mathrm{i}}}{(1/\beta) + \cos(\theta - \phi)\mathrm{d}p/\mathrm{d}\Delta}$$

t_i is the time interval in seconds from the initiation, obtained from the seismograms; β is the presumed velocity of rupture, dp/dΔ is slope of the travel time curve at delta, θ is the azimuth from the initial rupture to the recording station. The time lag of the subsequent events is governed by the azimuth and the distance of the recording station along with the position and the time difference of the later events with respect to the initiation. The second event was located at 6 km from the initiation at the CWPRS station and the 3rd event at 17 km, using Gutenberg's sine-curve approach. These are plotted in Fig. 6.3. These analyses supported the Koyna earthquake of 10 December 1967 being a multiple event earthquake. It is inferred that the rupture initiated at the CWPRS hypocenter and continued southward for about 6 s with a rupture velocity of 3.4 km/s. Langston (1976) estimated the source duration of 6.5 ± 1.5 s which is consistent with Gupta et al. (1971).

The earthquake focal depth estimates also varied as can be seen from Table 6.1. The estimate of a focal depth of 10 km by Singh et al. (1975) and 4.5 ± 1.5 by Langston (1976) appear to be most appropriate. The large felt area is due to efficient transmission of seismic energy in the Peninsular Shield region devoid of sediments.

The magnitude estimates also varied quite a bit by different agencies, as can be noted from Table 8.1. A surface wave magnitude of 6.2 was estimated by Narain and Gupta (1968) analyzing the long-period WWSSN seismograms of several observatories. Later, a more definitive surface wave magnitude of 6.3 was ascertained by Singh et al. (1975).

6.2 Strong Motion Data and Intensity

A study of the macroseismic effects of the Koyna earthquake revealed a very localized zone of high intensity corresponding to VIII intensity on the Modified Mercalli (MM) Intensity Scale. The long distances of up to 700 km where the earthquake was felt is attributed to efficient transmission of the seismic energy in the Indian Shield region devoid of a sedimentary cover. A generally accepted intensity distribution of the Koyna earthquake, adopted from Geological Survey of India (1968) is depicted in Fig. 6.6.

An interesting observation has been the documentation of the rotational movement in the meizoseismal area of the Koyna earthquake. Table 6.2 gives the rotational movement documented. The locations of the observation of the rotational movements, and the sense of rotation (clockwise or anticlockwise) is depicted in Fig. 6.3. Figure 6.7 depicts the rotational movement of 7°, anticlockwise documented at pillar at Donachiwada.

The horizontal component strong motion seismographs located in a gallery at about mid-height of the Koyna Dam are installed with orientation parallel and perpendicular to the Koyna Dam. The inferred particle motion from the accelerograms is consistent with the source propagating from north to south. Figure 6.8 depicts the accelerogram, adopted from Krishna et al. (1969), where the initiation and the second event of Koyna rupture are identified. The particle motion plotted in Fig. 6.8 depicts the first 11 s of the motion. That is also consistent with the north to south rupture propagation.

Very interesting field observations were made immediately after the occurrence of the Koyna earthquake at the dam site

Table 6.1 Focal parameters of the Koyna earthquake of December 10, 1967

Agency	Epicenter		Origin time (GMT)	Magnitude		Depth (km)
	Lat. (N)	Long. (E)		m_b	Ms	
Central Water and Power Research Station (C.W.P.R.S.)	17° 31.1′	73° 43.9′	22:51:17.0	7.0		12
India Meteorological Department (I.M.D.)	17° 22.4′	73° 44.8′	22:51:19.0	7.5		8
Moscow Academy of Sciences of the U.S.S.R. (M.O.S.)	17° 30′	73° 48′	22:51:19.0	6.4	6.5	
Bureau Central International de Seismologie (B.C.I.S.)	17° 24′	73° 36′	22:51:20.0	6.4		
International Seismological Summary (I.S.S.)	17° 32.4′	73° 50.4′	22:51:23.2	5.9		Depth from P excessively negative
United States Coast and Geodetic Survey (U.S.C.G.S.)	17° 39.6′	73° 55.8′	22:51:24.3	6.0		33 km restricted

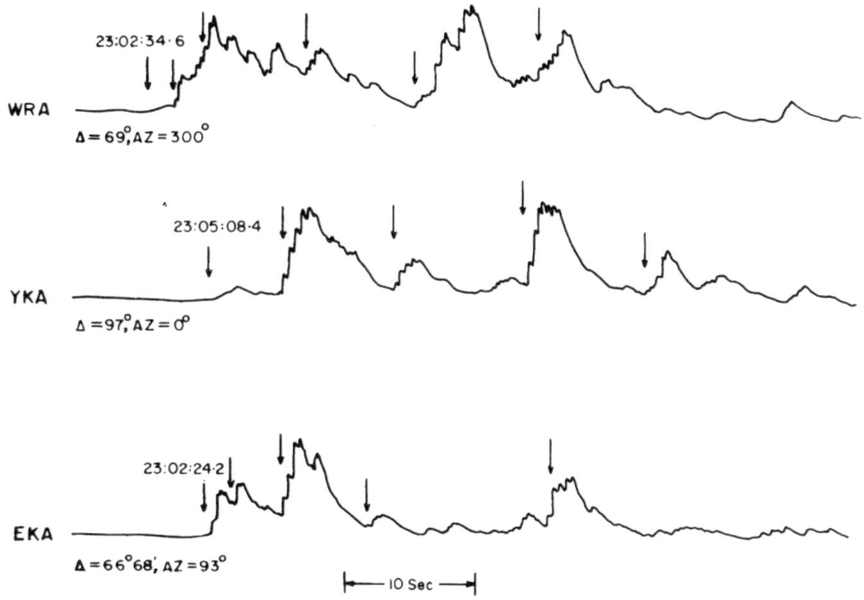

Fig. 6.1 Correlograms of the seismic arrays at Warramunga (WRA), Yellowknife (YKA) and Eksdalemuir (EKA). Six events are marked by arrows

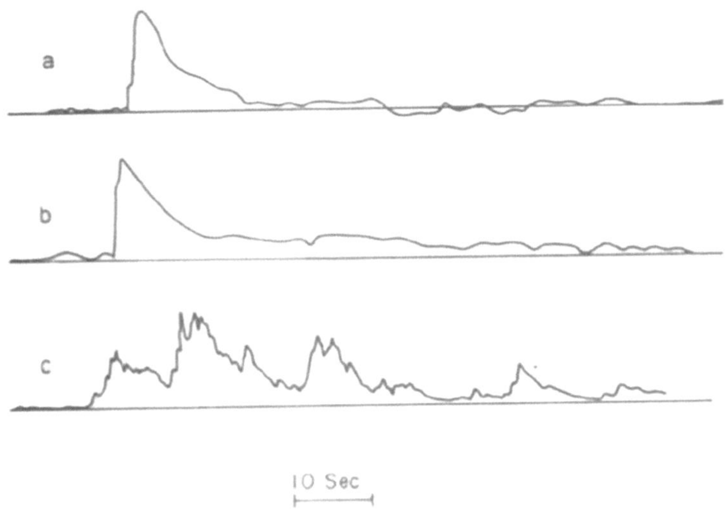

Fig. 6.2 Correlograms for an explosion (**a**), a simple earthquake (**b**), and a complex earthquake (**c**) (after Thirlaway 1963)

and nearby regions. A full description has been provided in Gupta and Rastogi (1976). Among the few very interesting observations are the movement of the unfixed tiles on the span of the Koyna Dam, demonstrating an increase in the displacement towards the center of the Koyna Dam (Fig. 6.9) and the compression documented at the center of the Koyna Dam (Fig. 6.10). Another interesting observation is the documentation of the compression at the center of the Koyna dam depicted by a bent steel ladder at the control panel room situated on monolith 24 (Fig. 6.11).

6.3 Major Geological Features of Koyna Region

A good description of the geology of the Koyna region has been provided by Krishnan (1960), Geological Survey of India (1968), and the Committee of Experts (1968) set up by the Government of India to study the Koyna earthquake.

A bit less than one-half of the Peninsular Shield of India, where the Koyna Dam is located, is covered by the Deccan Traps. Archean and Upper Precambrian formations are

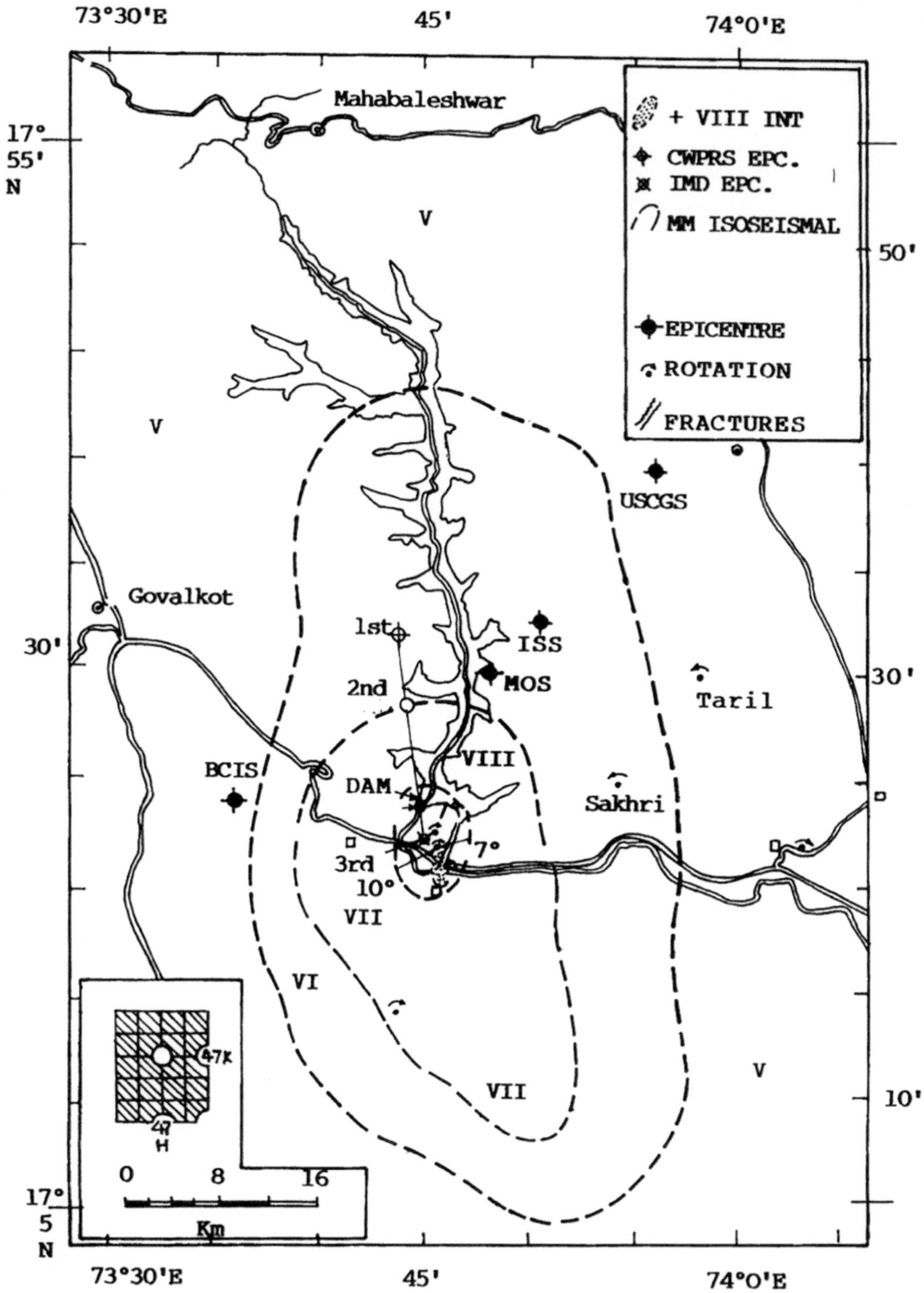

Fig. 6.3 Map of the Koyna region showing the seismic stations, the epicenters located by various agencies, isoseismals, and the rotational displacements. The later events, located by the analysis of the delay times, are shown as "1st", "2nd" and "3rd"

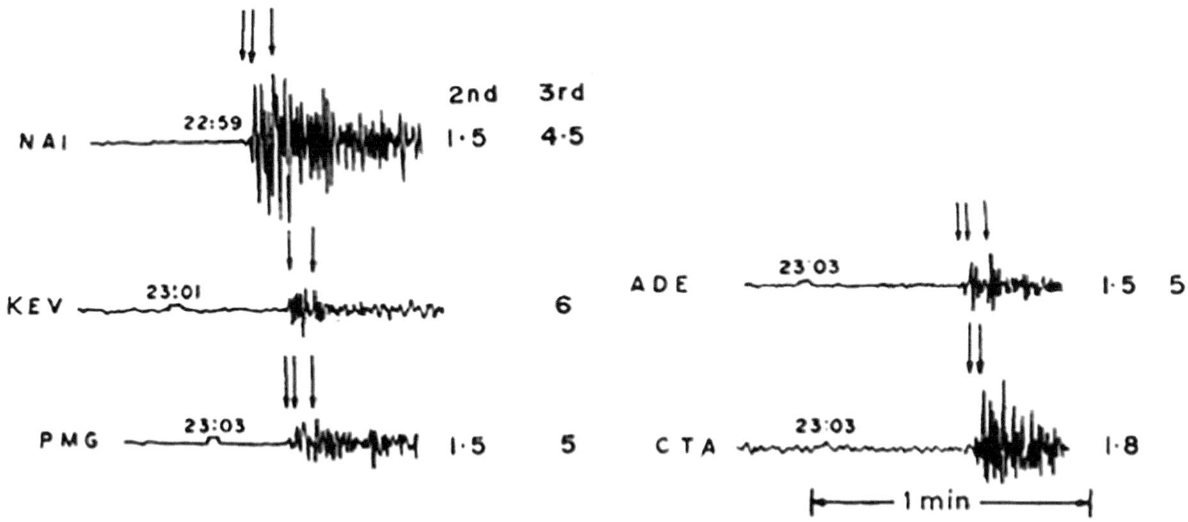

Fig. 6.4 P-wave portions of the short-period seismograms from a few stations. The initiation and later events are marked by arrows and the delay times (in seconds) are noted alongside

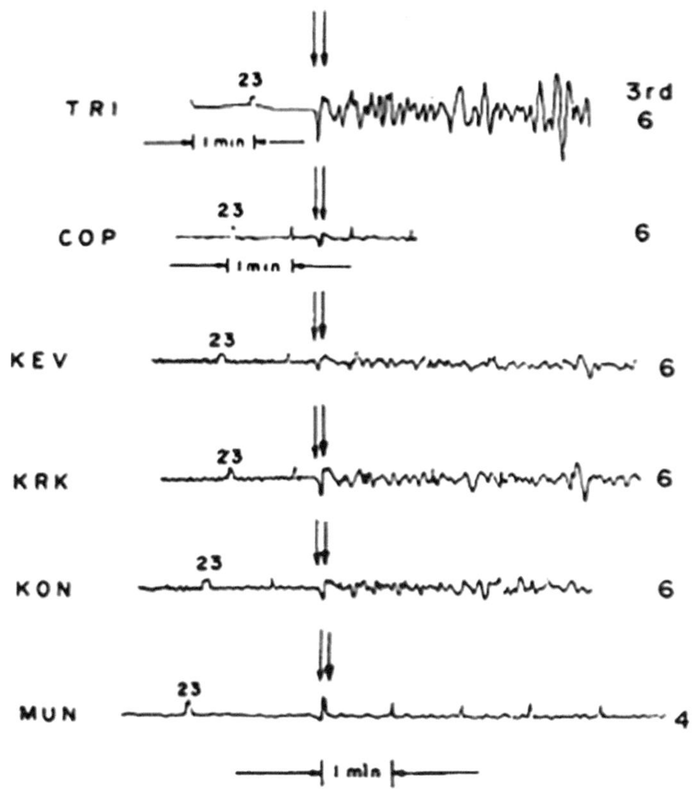

Fig. 6.5 Tracing of the P-wave portion of the long-period seismograms from some of the stations. The initiation and the third event are marked by arrows and the delay times (in seconds) are noted alongside

exposed over the rest of Indian Peninsular Shield. The major mountain building activity ceased in Precambrian times. However, minor folding, block faulting was observed later, including epeiorogenic movement. Figure 6.12 provides a simplified geological map of Peninsular India.

Basaltic flows erupted from numerous fractures in the crust during uppermost Cretaceous to Eocene times. These have been well described in 1960 by Krishnan (1960). One of the largest volcanic flood eruptions constitutes the Deccan Flood Basalt, covering an area of about half a million km^2

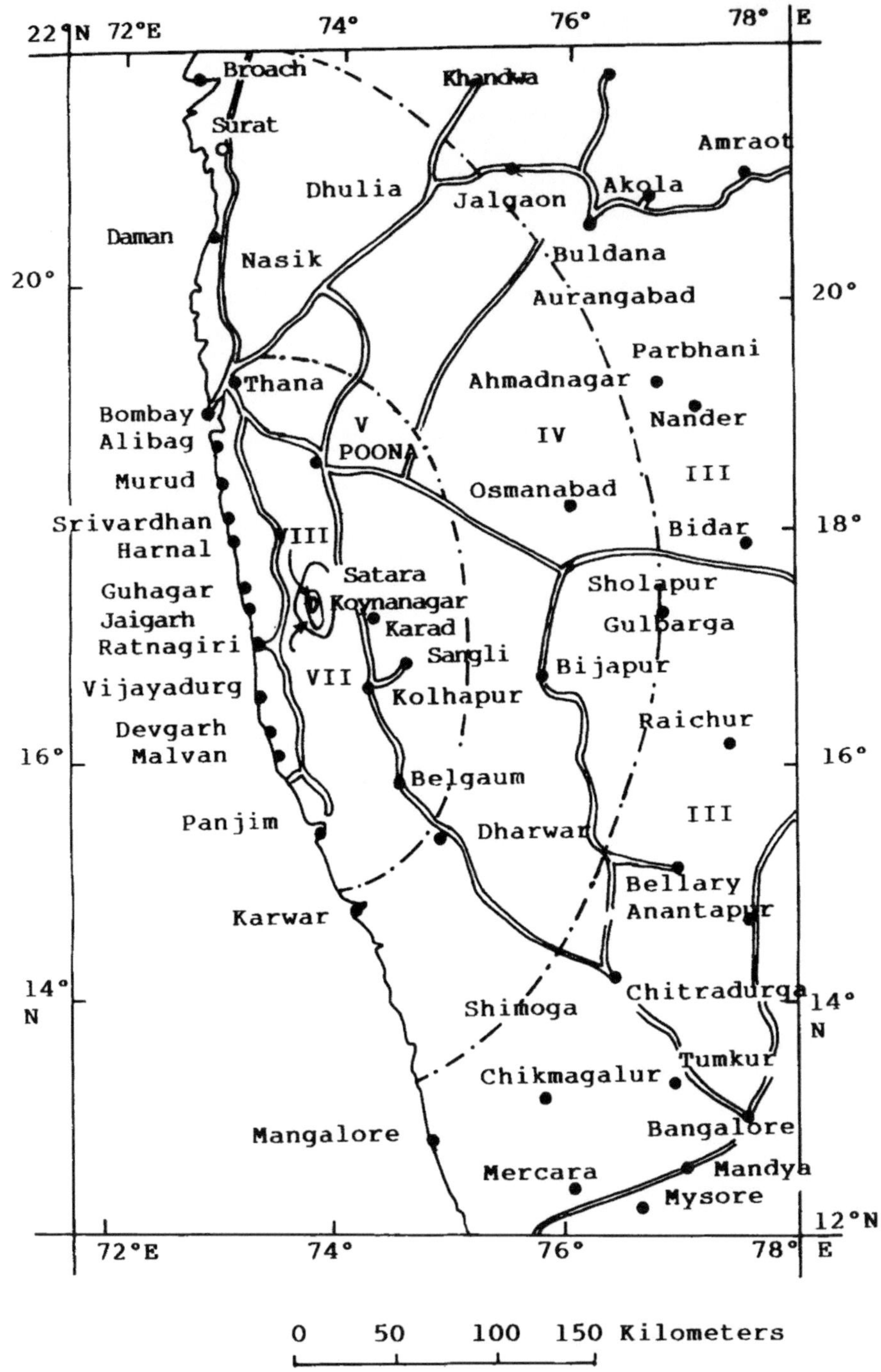

Fig. 6.6 Map showing the isoseismals in Western India (modified from the Geological Survey of India 1968)

on the land (Fig. 6.12). The basaltic lava flows are basically horizontal sheets. The initial flows filled up the irregular topography caused by erosion during the Late Mesozoic times. There is quite a bit of elevation difference, with the highest flows having an elevation of 800 m and descending to sea level in about 70 km distance. The name *Traps* is given due to their terrace like appearance. The maximum thickness of the flows is estimated to be 2 km in the Konkan and the Western Ghats region located east of Bombay and then thinning out further east to Nagpur where there is only about 100 m thickness of the traps. The Western Ghats escarpment exposes almost a 1300 m section of the flows.

Table 6.2 List of the locations where rotational movements have been measured with their magnitudes and directions

Location	Rotation	Recorded by
Sakhri	5° anticlockwise	Geological Survey of India
Tarli	12° anticlockwise	Geological Survey of India
Wai	14° anticlockwise	Geological Survey of India
Nanel	7° clockwise	Geological Survey of India
Donachiwada (N)	10° clockwise	Geological Survey of India
Chandol	5° clockwise	Geological Survey of India
Khelkarwadi	3° clockwise	Geological Survey of India
Donachiwada (S)	7° anticlockwise	National Geophysical Research Institute

Fig. 6.7 Rotational displacement of a pillar at Donachiwada (photograph by H. K. Gupta)

The Deccan Traps are mostly spread horizontally. A very gentle easterly dip of 1° is exhibited in the high plateau region. However, in the Konkan area, a westerly dip of 3–4° is seen (Das and Ray 1972). A dip of 15° is observed in the westerly direction called the Panvel Flexure due to monoclinal nature of the structure. Major faults have not been reported in Deccan Traps. There is a belief, with a few geologists that the Western Ghats that extend some 450 km in the north–south direction are a manifestation of a deep-seated fault (Committee of Experts 1968). This concept is supported by the 33 hot springs located between Surat and Ratnagiri. Geophysical investigations support the existence of north–south running satellite fault along the western bank of Koyna River. The observed straight west coast of India from

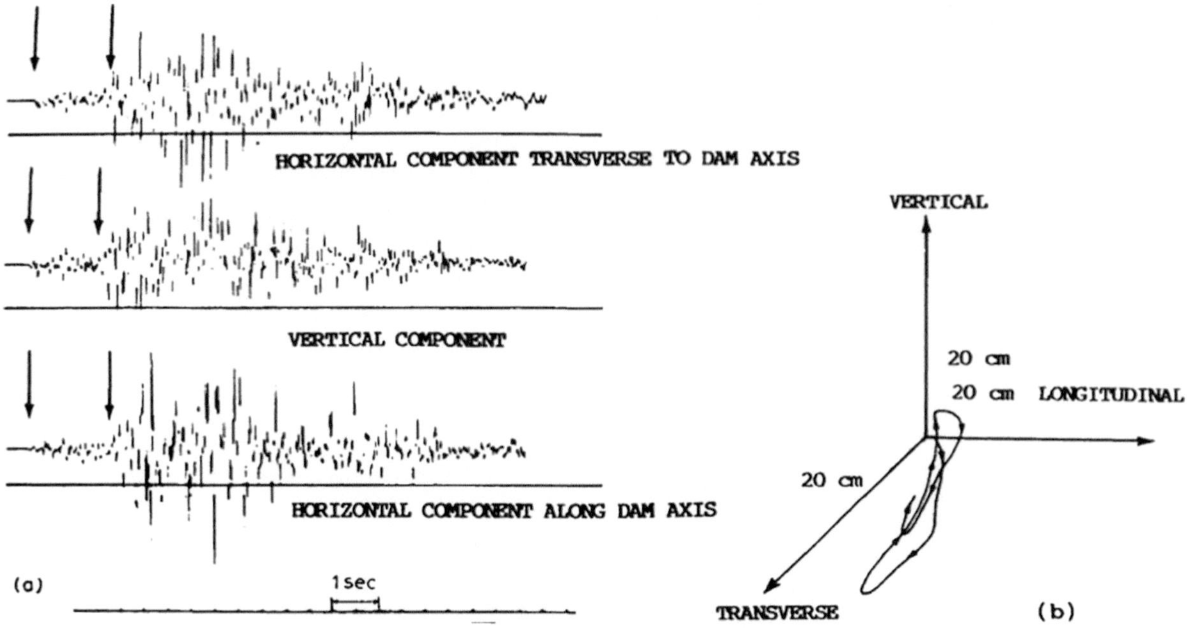

Fig. 6.8 **a** Accelerograms for the Koyna earthquake recorded at the dam (adapted from Krishna et al. 1969); the initiation and second event are marked by arrows. **b** Particle motion, obtained from the accelerograms, plotted for about 11 initial seconds

Fig. 6.9 Movement of the unfixed tiles on the span of the Koyna Dam. Note the increase in displacement towards the center of the span from α to c (photograph by H. K. Gupta)

Cambay to the southern tip of India is mostly attributed to a major fault during the Late Pliocene to Early Pleistocene time. During the Pliocene to the Recent times several epeiorogenic movements are reported along the west coast of India.

A series of fractures associated with brecciation, and shattering occurred in the Peninsular Shield of India. Evidence for displacement are missing. It is believed that the shearing of

Fig. 6.10 Compression documented at the center of the Koyna Dam span (photograph by H. K. Gupta)

the crust is mostly responsible for the fractures. Similar fractures are observed in Deccan Traps oriented in NNE–WNW direction. Das and Ray (1972) have inferred that the fractures along the Panvel Flexure, with NNW to NNE alignment, are shear and tension joints differing from the normal cooling joints observed in the basaltic lava flows. Sheared zone rocks exhibit shattered nature and slicken sides are observed indicating movement along shear planes. The observed fractures may have been most likely due to secondary stresses caused by the re-activation of the suspected basement fault. In the western and northern parts of the Koyna catchment area, the fractures are aligned in NNW–NNE direction, running some 20 km and cutting through 800 m of volcanics without any displacement. In the stage III of the Koyna Project, in the tail-race tunnel, a group of fractures with orientation from N to NNW is observed. In this part, the rocks are observed to be shattered forming clay gauges in the fracture zones, with the width varying from 1 to 20 m. It is inferred that the N–S stretch of 500–600 m deep Koyna Valley, which is against the general slope of the host country, may have been caused due to a fault or fractures. The sharp eastward turn of the North–South flowing Koyna River, as can be seen in Fig. 6.3, must have been caused by a major E-W fracture or a fault.

6.4 Initial Studies of Gravity Anomalies, Refraction Profiles and Hot Springs

A negative Bouguer anomaly prevails over the Peninsular India with an average value of – 60 milli gals. In the Koyna region, the anomaly increases to – 100 milli gals (Bhaskar Rao et al. 1969; Kailasam et al. 1972). This negative anomaly has been interpreted differently by geophysicists. It is attributed to a synclinal sag by Kailasam et al. (1972). Guha et al. (1974b) have attributed it to a buried anticline of low density constituted by low-density granite of Kaladgis, with a density of 2.7 compared to 2.9 of basalts. According to Brahmam and Negi (1973), it may have been caused by the rift valley that developed prior to the deposition during the Precambrian and possibly accommodating Precambrian formations and Gondwana sediments. Deep refraction soundings have been reported by Kailasam et al. (1969) at Karad, Pophli and Rampur. From these surveys discontinuities with a higher velocity layer at depths of 500 m near Karad, 200 m near Pophli, and 670 m near Rampur have been inferred. The existence of a fault near Rampur, as inferred by regional gravity traverses, is supported by these deep refraction soundings. This is consistent with

Fig. 6.11 Bent steel ladder at the control panel room situated on monolith 24 (photograph by H. K. Gupta)

the N–S trending zone of 33 hot springs located between Surat and Ratnagiri. Kailasam and Murthy (1969) report inferences drawn from other gravity/magnetic profiles run across the Koyna River south of the Koyna Dam, located just before the easterly turn of the river. On the western bank of the Koyna River a sharp drop in Bouguer gravity is very noticeable (Fig. 6.13). Kailasam and Murthy (1969), on the bases of these observations suggested the existence of N–S trending fault in the direction of the Koyna River course, which is hypothesized to extend to the basement.

6.5 Studies of Seismicity and Water Reservoir Levels at Koyna During 1962 Through 1990

6.5.1 Seismicity

The Koyna Dam is in the Peninsular India, which is one of the world's stable continental Precambrian Shields. The region is basically aseismic. In the several decades preceding the occurrence of the 10 December 1967 Koyna earthquake, no such

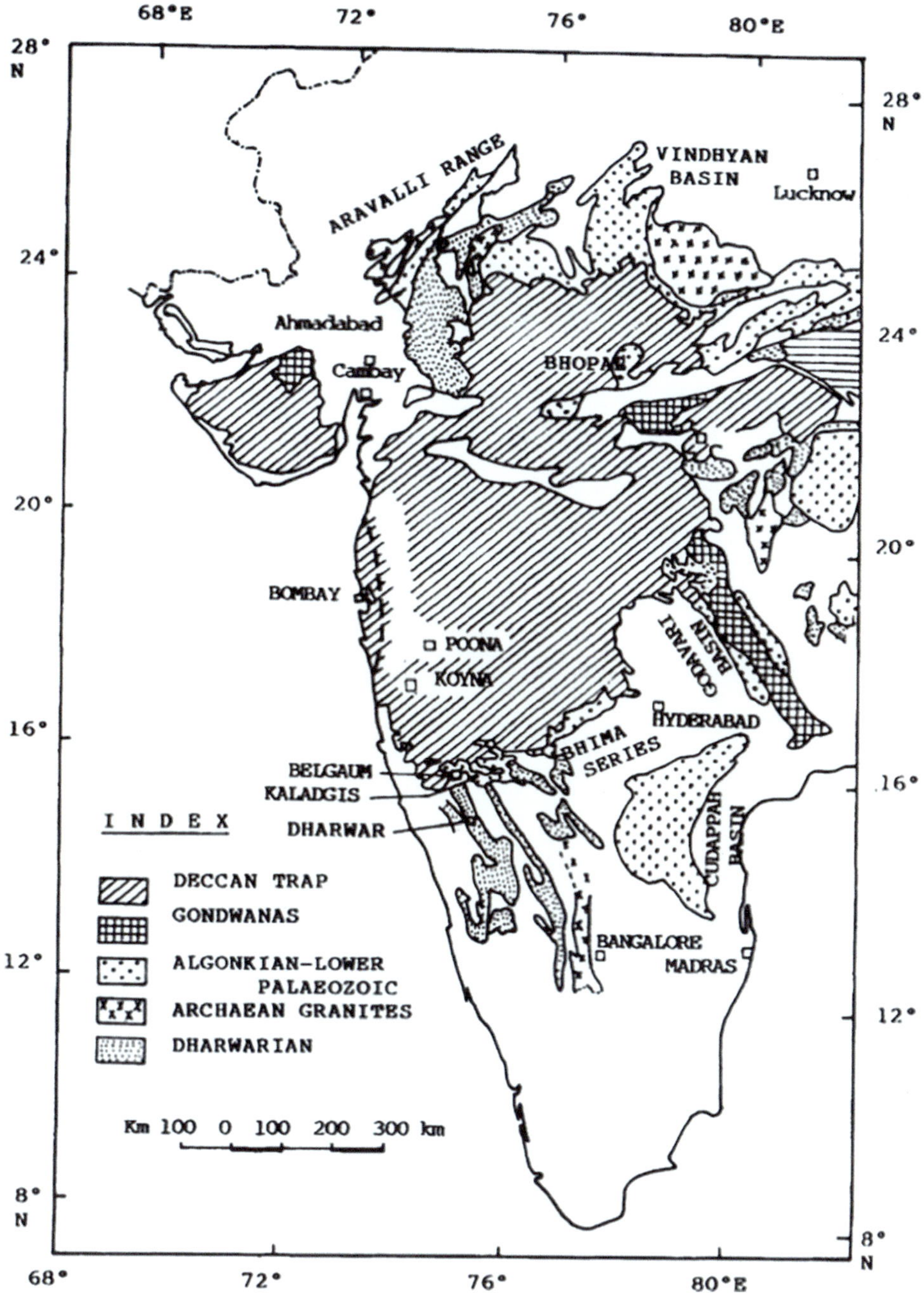

Fig. 6.12 Simplified geological map of Peninsular India

earthquake is known to have occurred. A scrutiny of the historical records during the past 600 years, however, reveals occurrence of mild tremors, particularly along the west coast of India (De Silter 1965). Sykes (1970) suggested that the earthquake occurrence along the west coast of India is possibly due to the postulated extension of the trench in the Indian Ocean between Ceylon and India. A catalogue of earthquakes prepared by Guha et al. (1970) for the period 1341 through 1969 has reported the 1843 Bellary and 1900 Coimbatore earthquakes. However, Coimbatore is located some 600 km away from Koyna and Bellary is some 500 km from Koyna. No seismic station operated in the vicinity of Shivaji Sagar Lake created by the construction of the Koyna Dam. Many aged residents within 50 km radius of the Koyna Dam, when interviewed, categorically denied of having experienced any tremor in the region prior to the impoundment of the Koyna Dam in 1962. An examination of the seismograms of Benioff seismometers operating at the Poona Seismological Station, 115 km from Koyna, since 1950, also did not show any tremors occurring in the vicinity of the Shivaji Sagar Lake.

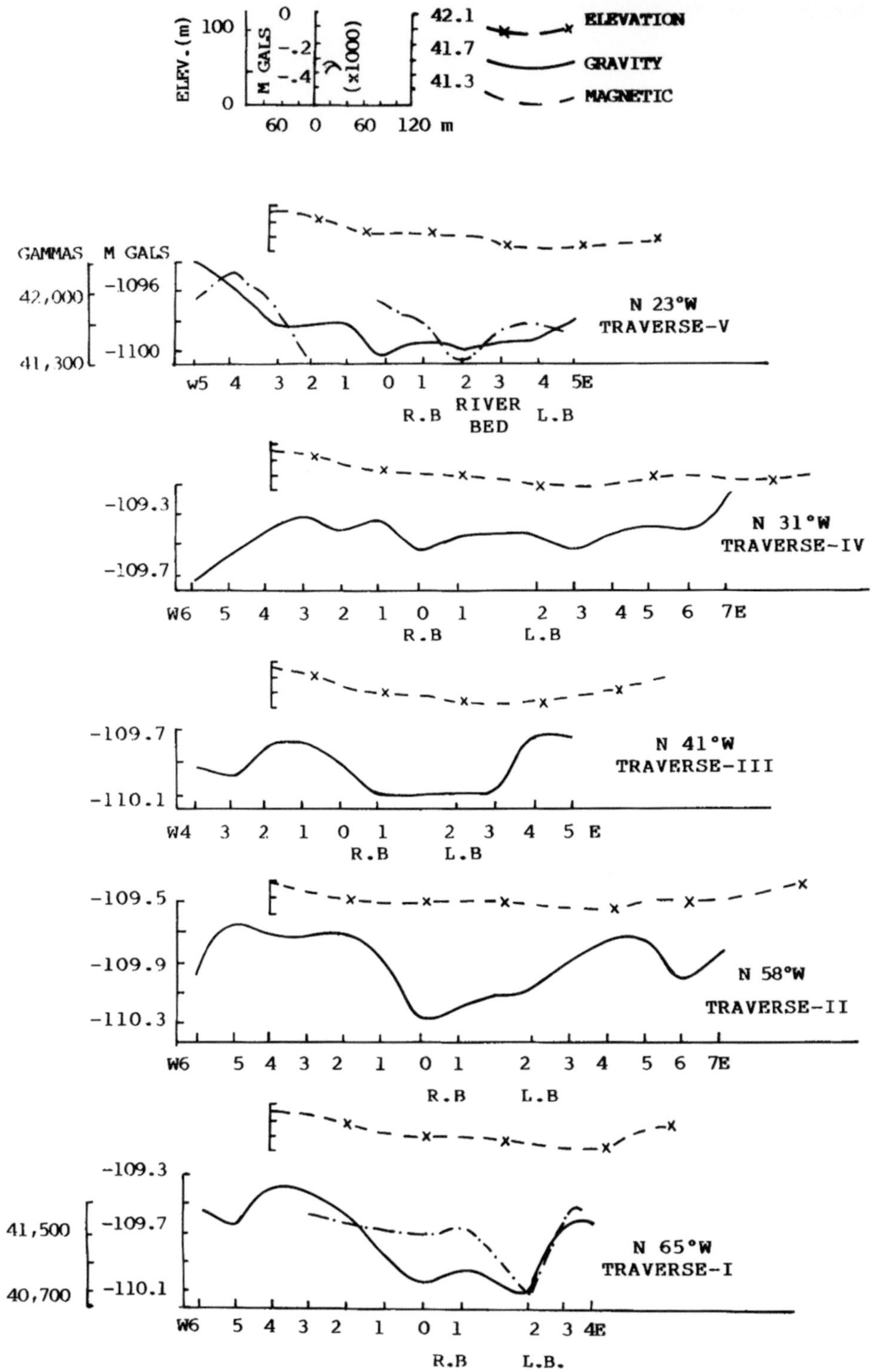

Fig. 6.13 Bouguer gravity anomalies and magnetic anomalies along the profiles across the Koyna River (after Kailasam and Murthy 1969)

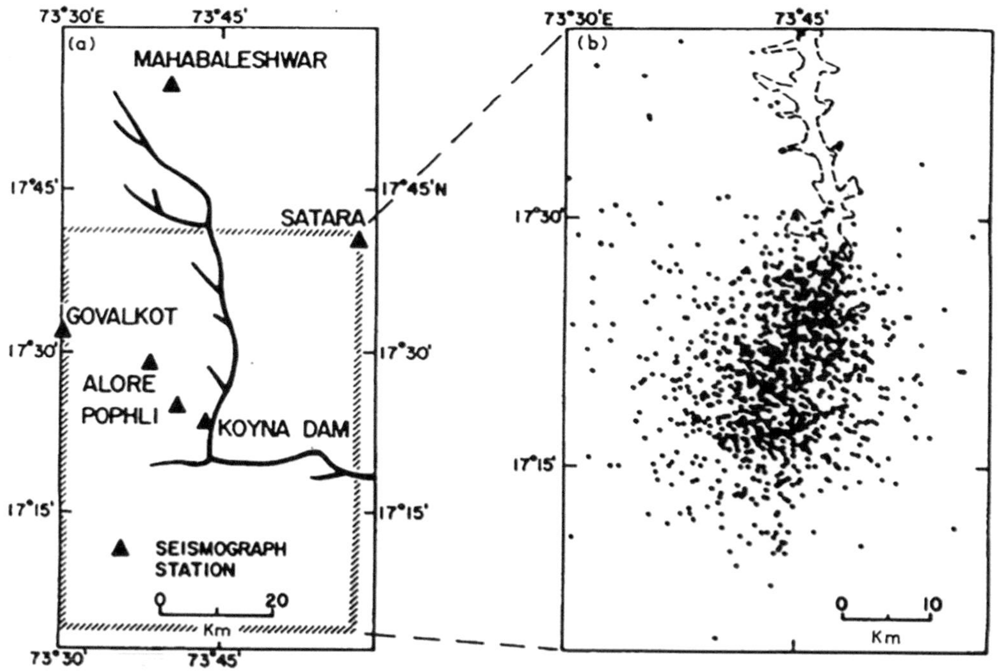

Fig. 6.14 **a** Seismic stations in the Koyna region, **b** epicenters located in the Koyna region for the period December 1967 through December 1971 by Guha et al. (1974b). Adapted from Gupta et al. (1980)

The filling of the reservoir commenced in 1962, and occurrence of mild tremors accompanied with sound became prevalent. From the middle of 1963, the intensity of tremors increased. The strongest of these tremors would rattle the windows. An intensity of V on the Modified Mercalli (MM) Scale could be assigned to these tremors. Installation of seismographs in the vicinity of the Koyna dam commenced by the end of 1963. A total of 7 seismic stations were established by 1969. This network included two stations in the immediate vicinity of the Koyna Dam (Fig. 6.14). The Koyna Dam observatories were also equipped with accelerographs and tiltmeters. Using the data from this network, Guha et al. (1968) determined the earthquake parameters. Figure 6.14b depicts the epicenters located for the period December 1967 through December 1971 in the vicinity of the Koyna Dam.

During the period October 1973 and December 1976, twelve earthquakes of Ms $\geq$ 4.0 occurred in the Koyna region. A detailed study of these 12 earthquakes, their foreshocks and aftershocks were carried out by Gupta et al (1980). They also developed the following relation for the surface wave magnitude of the earthquakes using the signal duration:

$$\text{Ms} = -2.44 + 2.61 \log t$$

t being the signal duration.

Figure 6.15 depicts the location of the epicenters as determined by Gupta et al. (1980). A north–south trending fault at 73° 45′ E could be identified with several of the epicenters located on this longitude, including the 10 December 1967 Koyna earthquake of M 6.3. The epicenter of the 8 December 1993 Ms 5.2 earthquake is also plotted in Fig. 6.15. It may be noted that 8 December 1993 earthquake is mostly attributed to the filling of the Warna reservoir in 1985. However, it is quite clear from Fig. 6.15 that the epicenter lies on the southward extension of the Koyna Fault.

6.6 Reservoir Levels and Triggering of Earthquakes

Gupta et al. (1969) drew attention to the water levels in the Shivaji Sagar Lake and the occurrence of earthquakes. From the limited data then available, it was pointed out that the highest water levels for the longest duration were retained in the Shivaji Sagar Lake from August to December 1967. This is well correlated with the most intense seismic activity observed in the region from September to December 1967 (Fig. 6.16). It may be noted from Fig. 6.16 that during the years 1964–1966 the rate of loading as well as the highest water levels reached were much lower than in 1967, and the correspondingly fewer triggered earthquakes occurred. Generally, the seismic activity follows the loading of the reservoir with certain lag of time.

In a later study, Gupta (1983) demonstrated a close relationship between the reservoir loading rates, the highest water levels reached and the duration of retention of high levels and the triggered seismicity. Two earthquakes exceeding M 5 occurred in the vicinity of the Koyna Dam in 1967, one in 1973, and 3 in 1980.

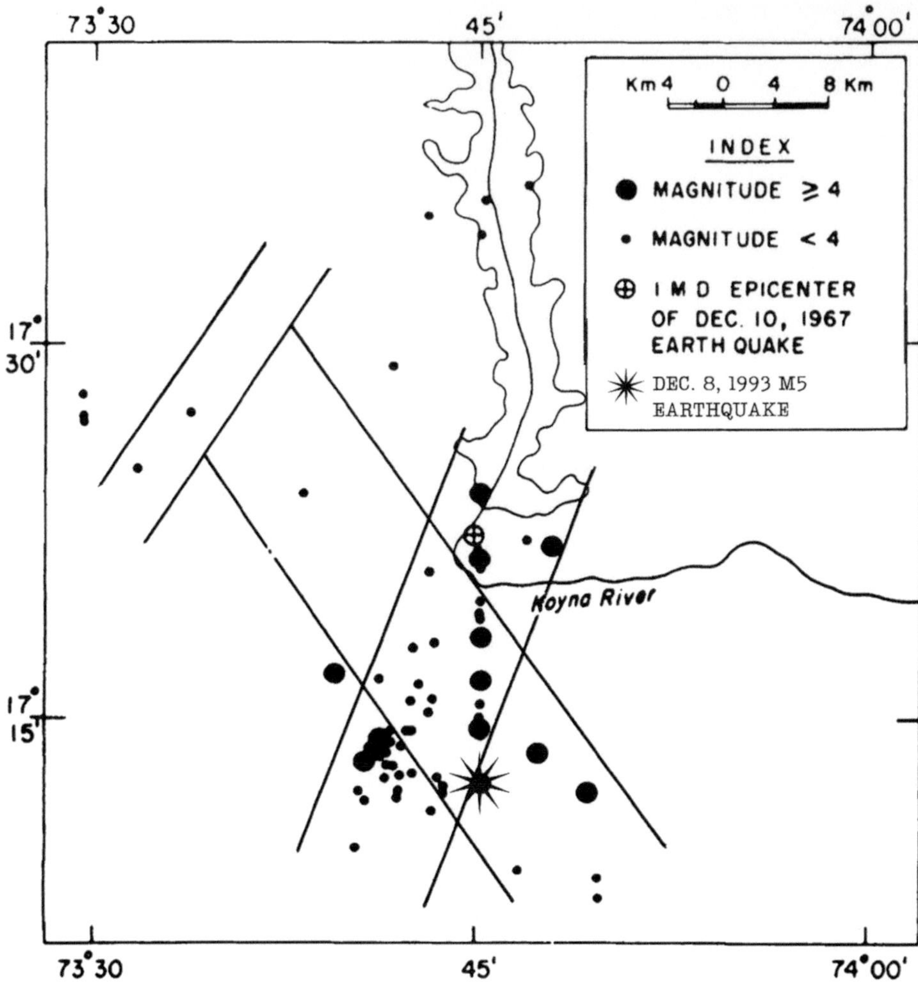

Fig. 6.15 Epicenters for the period 1973 through 1976 for M $\geq$ 4 earthquakes, their foreshocks and aftershocks located by Gupta et al. (1980). The three trends, from Rastogi and Talwani (1980), are shown by parallel lines. The epicenter of 8 December 1993 M 5.3 earthquake located at 17° 12′ N, 73° 45′ E lies on the extension of the north–south trending Donachiwada fault shown as a *

Figure 6.17, adopted from Gupta (1983), depicts the weekly changes in water levels at Shivaji Sagar Lake, and earthquakes of M $\geq$ 4.0 for the period 1975 through 1980. The highest water levels reached have exceeded 2160 ft level. During 1975 through 1980, the rate of loading has been less than 40 ft/week except for 1980. Three earthquakes of Ms $\geq$ 5.0 occurred in September 1980 (one on the 2nd of September, and two on 20th September). In the earlier years also, earthquakes of Ms $\geq$ 5.0 occurred when the loading rate exceeded 40 ft/week (Fig. 6.18). It may also be noted from Fig. 6.19 that in the years 1969 and 1971, the loading rate exceeded 40 ft/week, but no Ms $\geq$ 5 earthquake occurred. Gupta (1983) pointed out that a loading of 40 ft/week is a necessary but not a sufficient condition for Ms $\geq$ 5.0 earthquakes to occur in the Koyna region. The above work led to the conclusion that through proper monitoring of the loading of the Shivaji Sagar Lake, the occurrence of Ms $\geq$ 5.0 earthquakes in the Koyna region can be avoided. Simpson and Negmatullaev (1978) have reported similar findings for the Nurek Dam in Tajikistan.

It is concluded that smooth filling/emptying of the artificial water reservoirs can be useful in avoiding damaging reservoir triggered seismicity. The threshold rate of loading/unloading of a reservoir will vary from one location to another, and in the same location also it may vary with the passage of time.

6.7 Observation of Kaiser Effect at Koyna

Kaiser (1953), for the first time, noted that the acoustic emission under monotonically increasing stress, is found to increase when the applied stress exceeds the previously applied maximum elastic stress. This approach was successfully utilized by Yoshikawa and Mogi (1981) to estimate crustal stresses from the cored rock samples. The correspondence of the observed RTS at Nurek Dam and its relationship with the height of the water column in the reservoir has been noted by Simpson and Negmatullaev (1981). They found enhanced RTS when the water level in the Nurek Dam

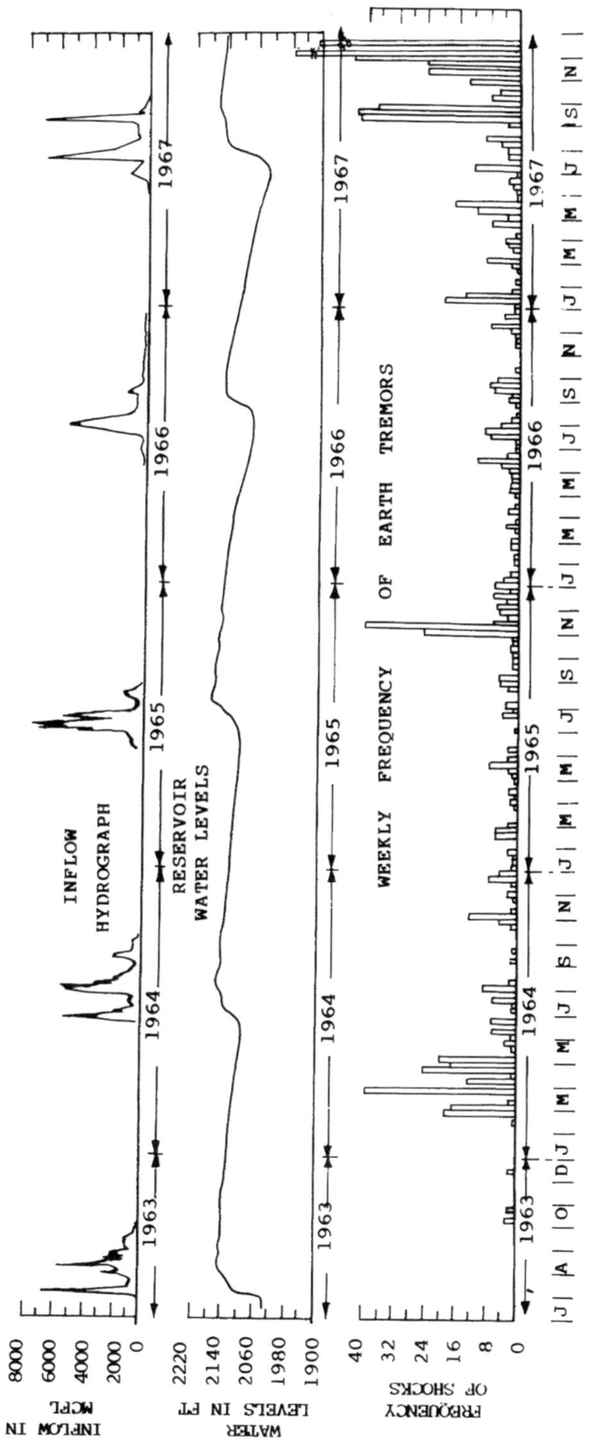

Fig. 6.16 Inflow hydrograph and water level of the Koyna Reservoir compared to the seismic activity in the area (after Guha et al. 1968)

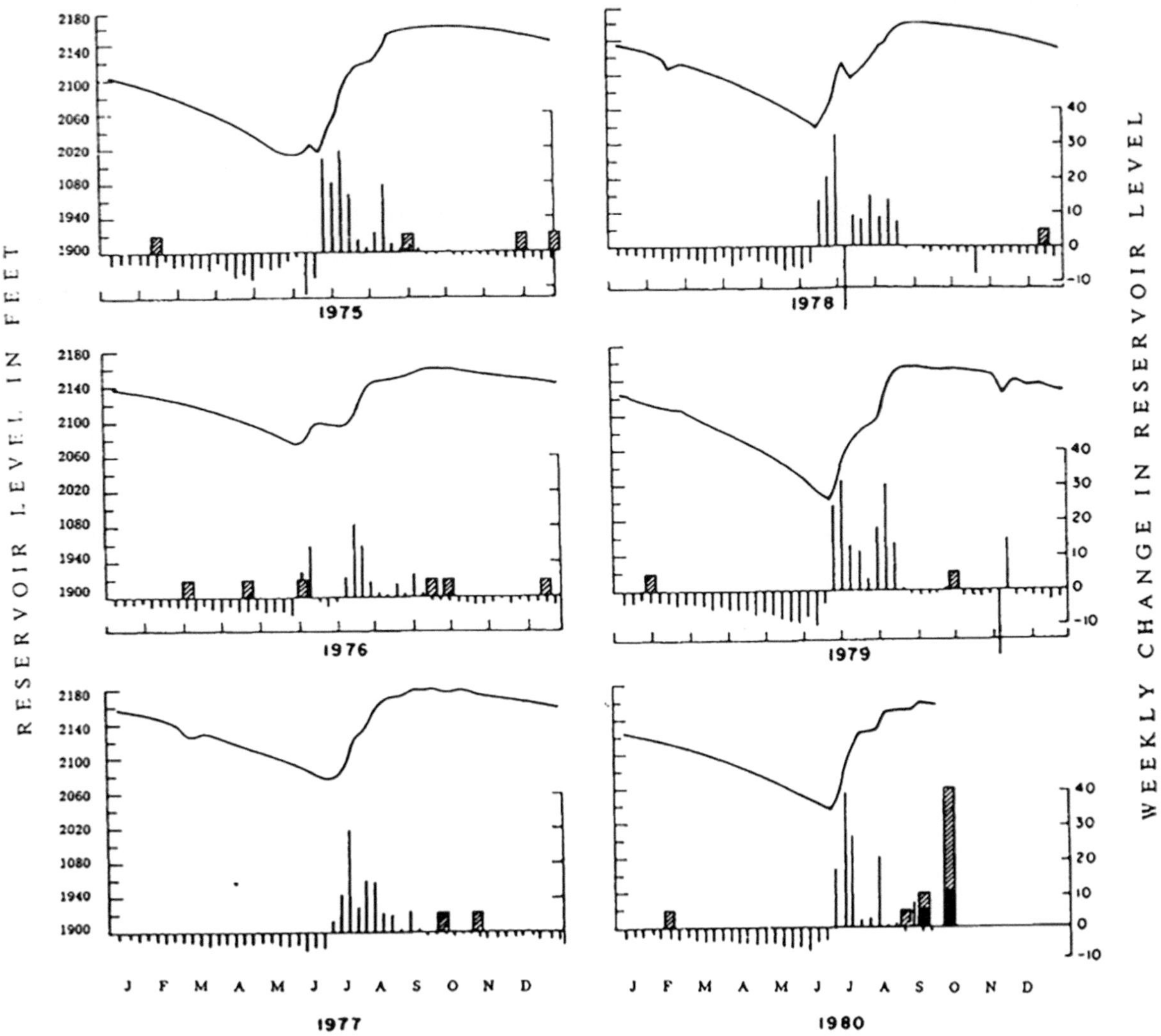

Fig. 6.17 Water levels at the Shivaji Sagar Lake (curve, scale on the left), weekly change in water levels (vertical bars, scale on the right), earthquakes of magnitude 4.0–4.9 (hatched column), M $\geq$ 5 (filled column). The height of the column is proportional to the number of earthquakes (e.g., one earthquake of magnitude $\geq$ 4 occurred on February 10, 1975, two earthquakes of magnitude $\geq$ 5 and six earthquakes of magnitude $\geq$ 4 occurred on September 20, 1980). Figure adapted from Gupta (1983)

exceeded the previous maxima. For Koyna it was observed that earthquakes of M $\geq$ 5 occurred in 1973 and 1980 within a month of the water levels in the Shivaji Sagar exceeding the previous maxima, as reported by Gupta and Combs (1976) and Talwani et al. (1996). Filling of the Warna Reservoir located some 30 km south of Koyna started in 1985. A water depth of 60 m at Warna was reached and Kaiser effect was observed in RTS in Koyna-Warna region during 1993 and 1994 (Talwani 1997).

6.8 Relocated Epicenters of M $\geq$ 4 in the Koyna Region

Figure 6.20 from Talwani (1997) shows the relocated epicenters of M $\geq$ 4 for the period 1974 through 1982. This demonstrates that RTS events were occurring south of the Koyna Reservoir up to 17° 10′ N latitude in the immediate vicinity of the Warna reservoir that was impounded in 1985.

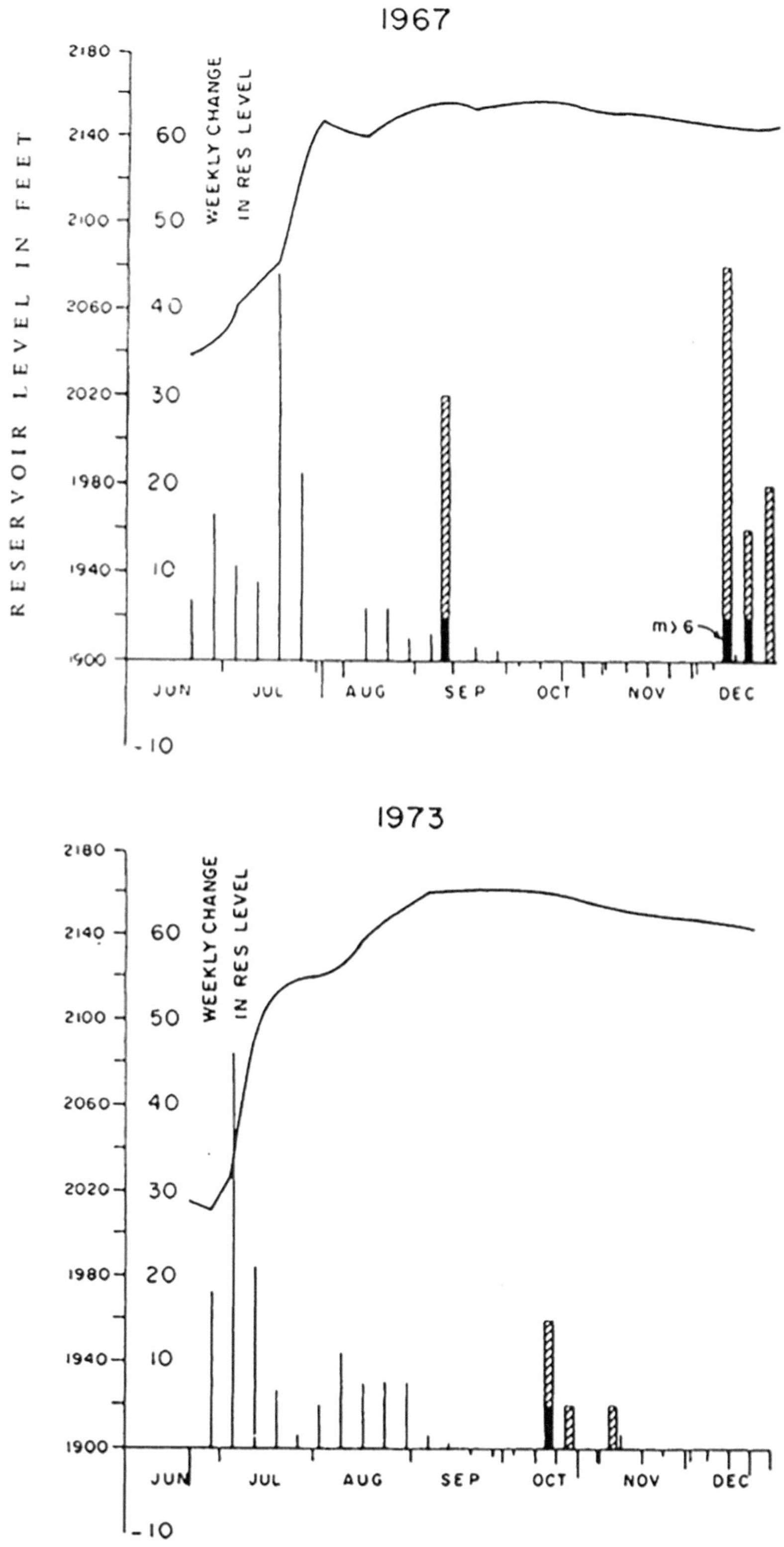

Fig. 6.18 The scheme of the figure is the same as for Fig. 6.17. The December 10, 1967 earthquake had a magnitude 6.3. Figure adapted from Gupta (1983)

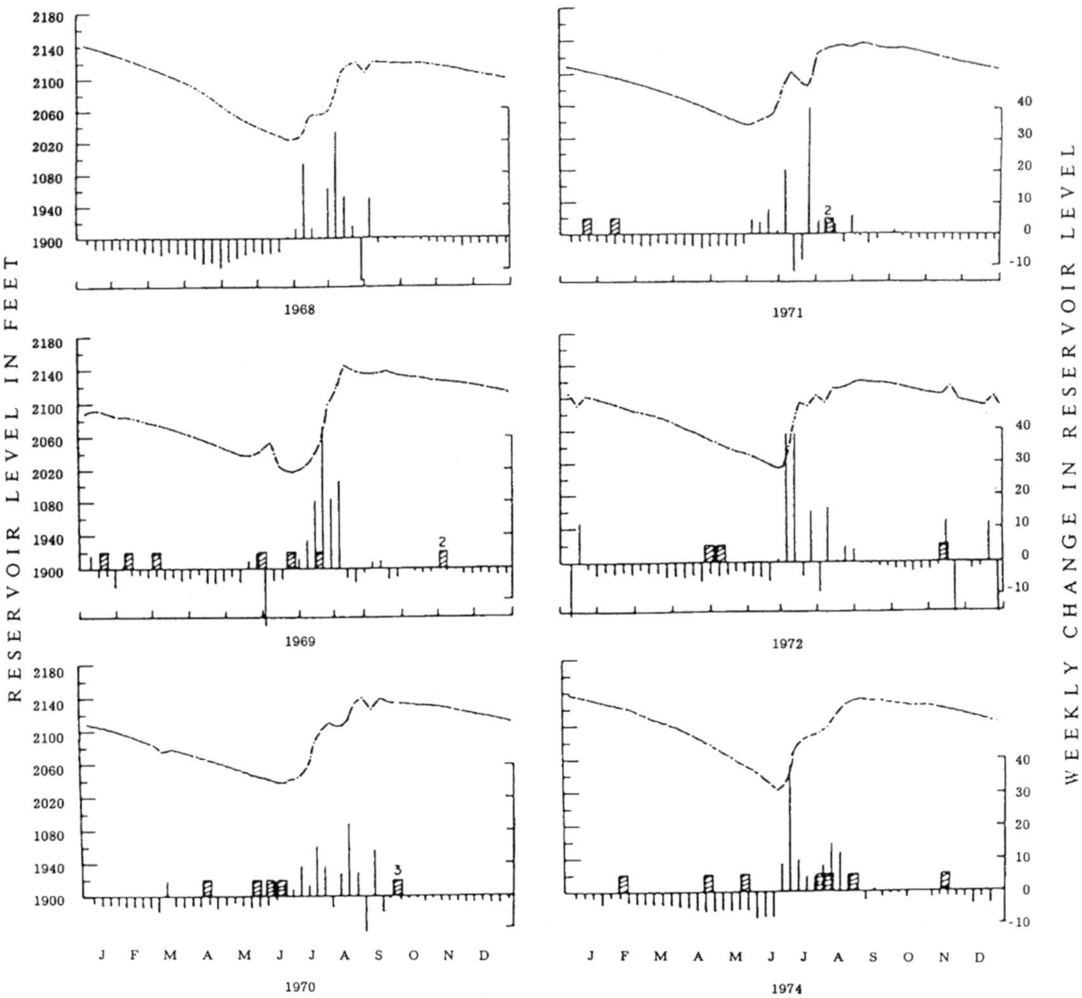

Fig. 6.19 Occurrence of more than one earthquake of magnitude > 4.0 is indicated by a number (e.g., three earthquakes of magnitude > 4.0 in one week of September 1970). Earthquake data are not shown for 1968 since there were numerous aftershocks of magnitude > 4.0 of the 10th of December 1967 earthquake. After Gupta (1983)

6.9 Migration of Foreshocks

An interesting phenomenon of the migration of foreshocks from shallow depths of 2–3 km to a depth of about 10 km has been observed for M $\geq$ 5 earthquakes in the Koyna region. Figure 6.21 depicts the seismic stations that were being operated in the Koyna region during 1993–95. During August 1993 through December 1995, a total of 1272 shocks of M $\geq$ 2 were recorded in the Koyna region. Figure 6.22 displays the foreshock-aftershock pattern for the 5.0 magnitude earthquake on December 8, 1993, along with the foreshock migration 400 h before the event. It may be noted that with the passage of time the depth of foreshocks increased from about 1 to 10 km. Similar patterns of foreshocks deepening to the depth of the main shock has been observed for several other earthquakes in the Koyna region.

6.10 Pore Pressure Studies

Fluctuations in water levels in the wells located in the vicinity of the earthquake epicenters have been reported globally. These could be seen before, during or after the earthquake occurrence (for example Roeloffs 1988). The fluctuations may represent pore pressure changes associated with redistribution of the stress in near and far field

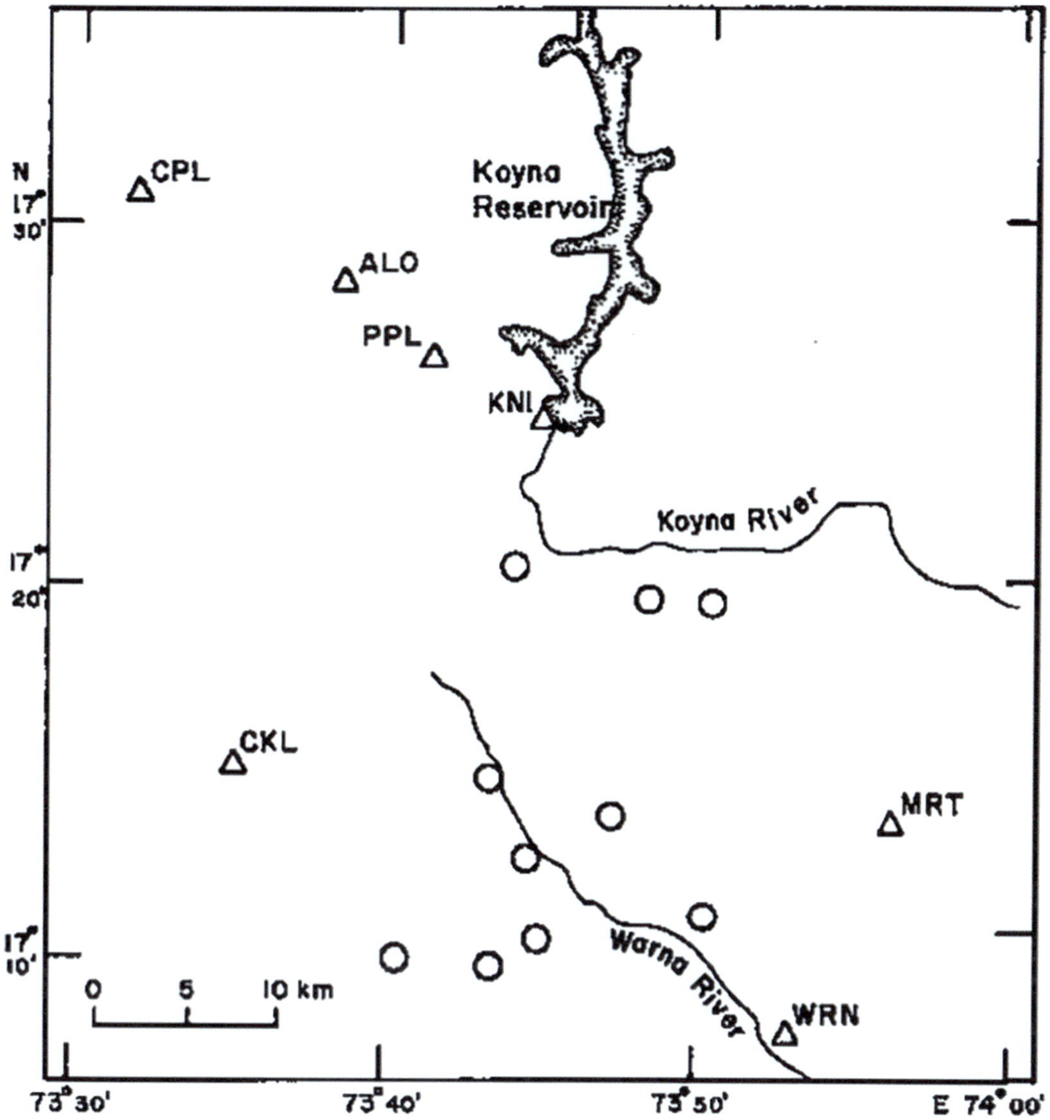

Fig. 6.20 Relocated epicentral locations of M ≥ 4 Koyna earthquakes between 1974 and 1982 (after Talwani 1997)

from the earthquake hypocenter. In collaboration with the University of Bonn, Germany, the National Geophysical Research Institute, Hyderabad set up a network of 21 borewells with depths varying from 90 to 250 m as depicted in Fig. 6.23. Most of the wells exhibited tidal signals, confirming the sensitivity of the wells to small strain changes (Chadha et al. 1997; Grecksch et al. 1999; Gupta et al 2000). This leads to an understanding that the wells shall also be sensitive to the water level changes in the Koyna and Warna reservoirs. The monitoring was on a continuous mode. It was the first experiment of monitoring well levels in an area of intense micro to macroseismic activity. A change of 2.5–6.0 cm was observed associated with the M 4.4 earthquake that occurred on 25 April 1997 as depicted in Fig. 6.24.

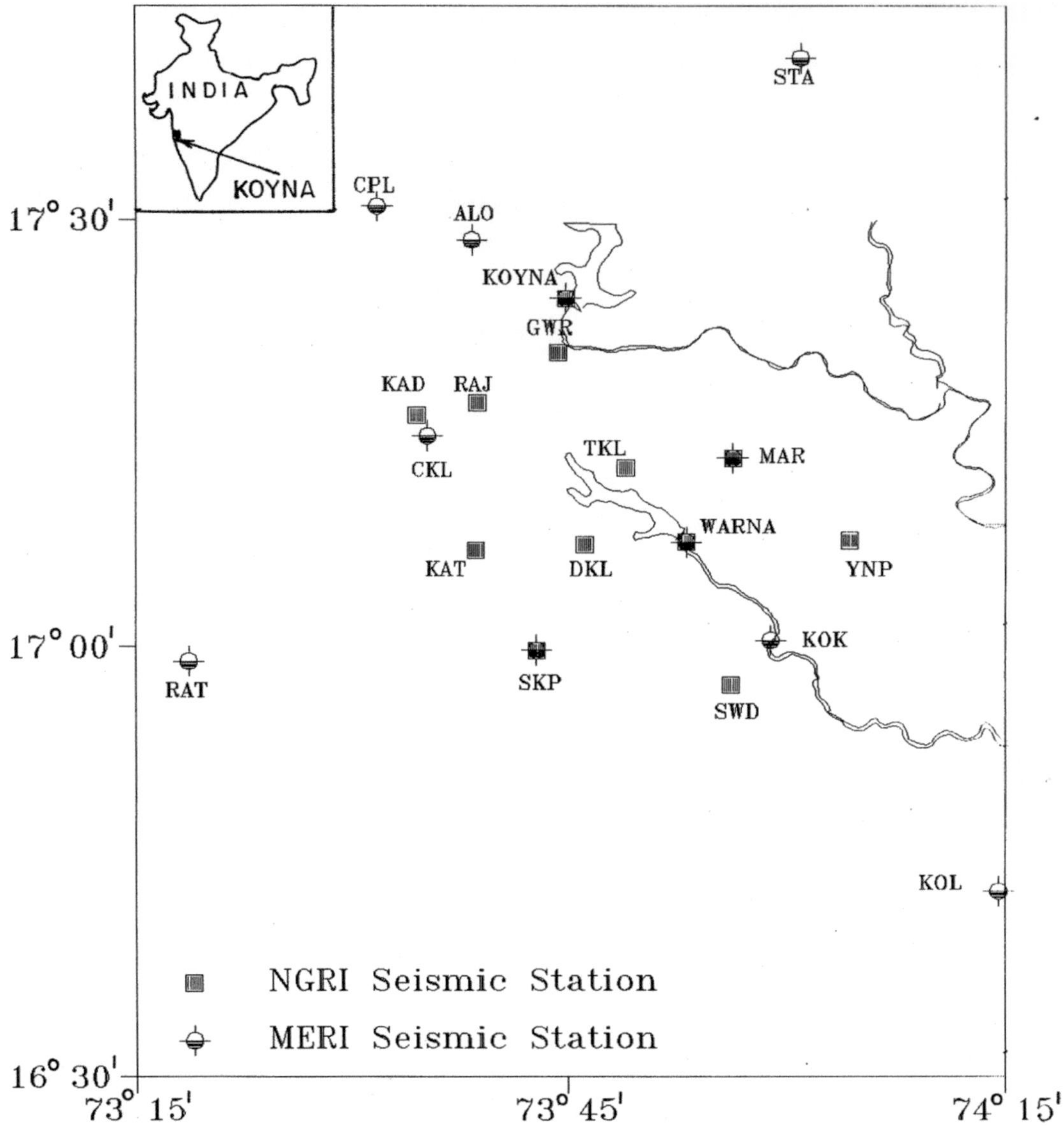

Fig. 6.21 Seismic stations operated by the National Geophysical Research Institute (NGRI), Hyderabad and Maharashtra Engineering Research Institute (MERI), Nasik in Koyna–Warna region during 1993 through 1995. The full length of the Shivaji Sagar Lake extending for about 50 km towards the north from Koyna is not shown

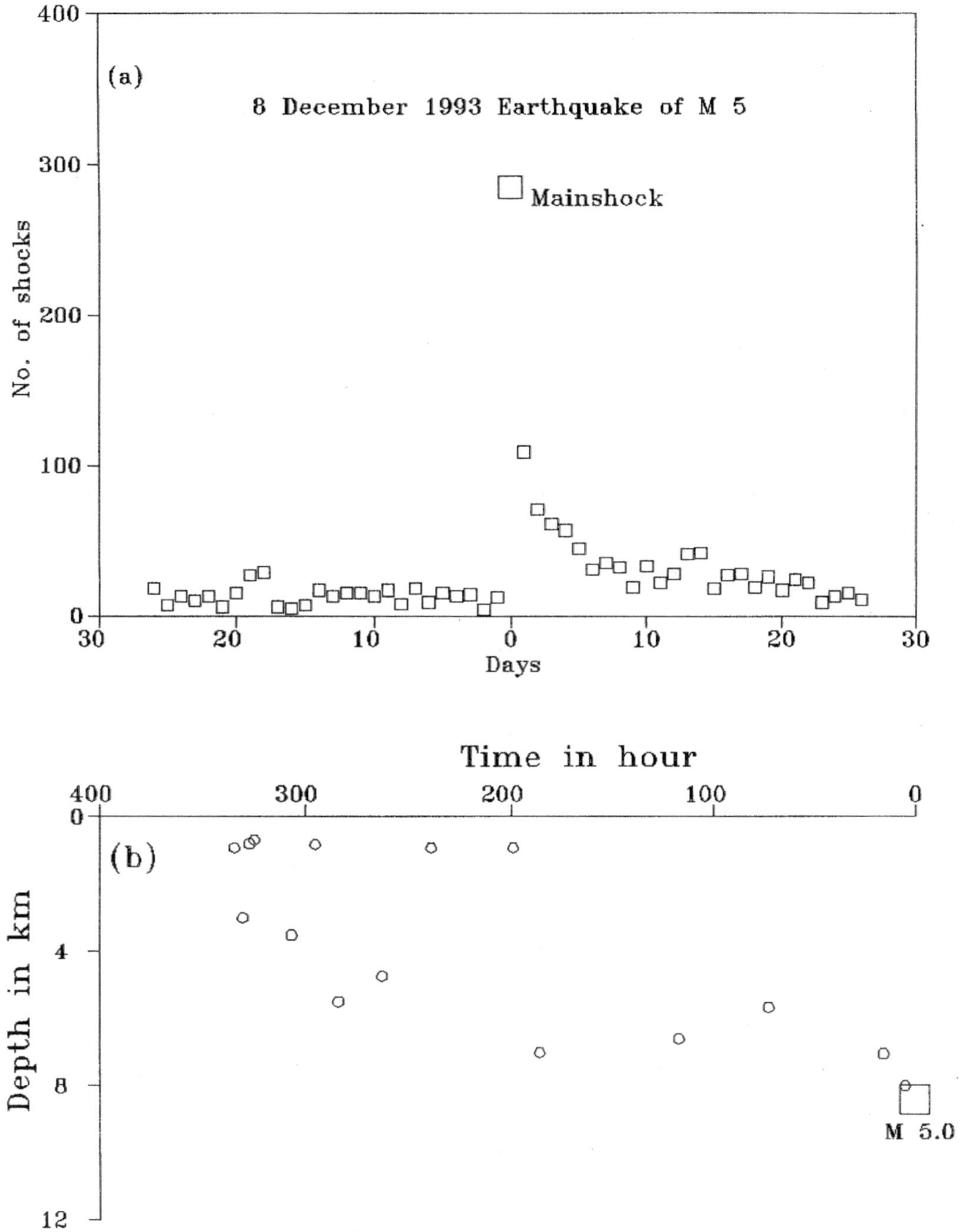

Fig. 6.22 **a** Foreshock-aftershock pattern of 8 December 1993 earthquake of *M* 5.0 close to Warna reservoir showing Mogi Type II pattern. **b** Temporal distribution of foreshocks of 8 December 1993 earthquake of *M* 5.0 with depth. Foreshocks have migrated from shallow depths (2–3 km) to the hypocenter of the main shock at 10 km, approximately 400 h before main shock

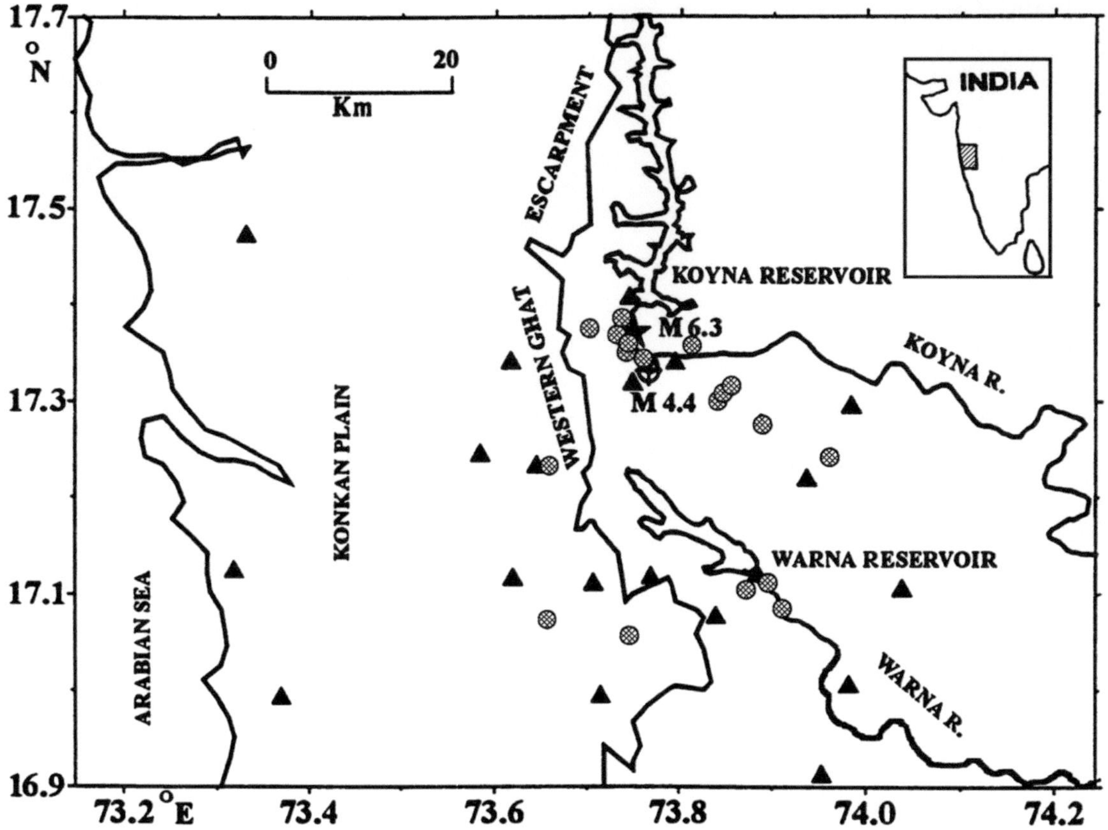

Fig. 6.23 The location of Koyna–Warna reservoirs in west India. Triangles indicate the sites where seismic stations are operated. Circles show the locations of borewells. The largest triggered earthquake of M 6.3 on 10 December 1967 is shown by a star. A circle and a plus sign show the M 4.4 earthquake of April 25, 1997, for which the residual pore pressure fluctuations are included

6.11 Soil-Helium Surveys in the Vicinity of the Donachiwada Fault Zone

In an interesting study soil-helium surveys were conducted in the vicinity of the Donachiwada Fault that hosted the 1967 M 6.3 as well as several earthquakes of M ~ 5 (Gupta et al. 1999). The Koyna earthquake occurred in the geomorphologically young Western Ghats terrain indicated by the almost straight course of the Koyna River before taking a sharp easterly turn. The NNE–SSW striking fault has been designated as the Donachiwada Fault by Harpster et al. (1979). Study of fault plane solutions of the earthquakes associated with the Donachiwada Fault have shown a strike-slip nature of the seismic fault with left-lateral movement (Talwani 1997). A soil-helium survey was conducted across the fissure zones along 12 east–west running profiles (Fig. 6.25). Along these 12 profiles the soil-gas was sampled at 5 m intervals at a depth of 1.6 m. The peak values along these observation points exceed from 1 to 7 ppm above atmospheric abundance of 5.24 (Fig. 6.26). Taking $^{3}He/^{4}He$ for air (R_{air}) as 1.0, the ratios in the soil-gas samples vary from 0.590 ± 0.010 to 0.632 ± 0.008, indicating

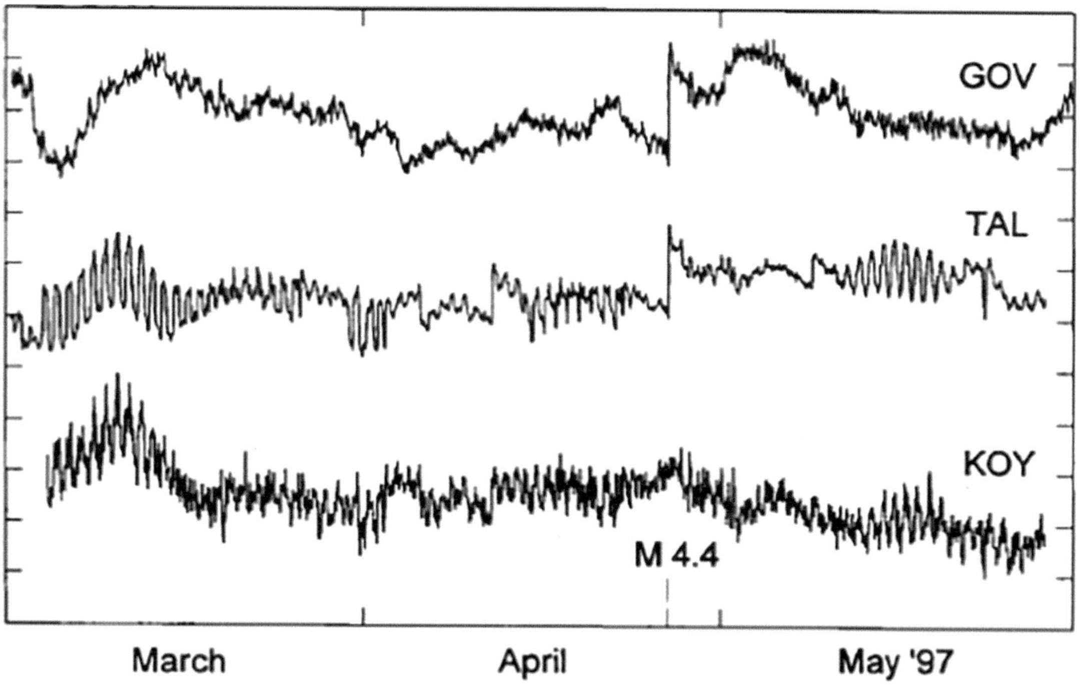

Fig. 6.24 Residual pore pressure fluctuations as observed in three adjacent wells. Co-seismic anomalies were observed at well sites GOV and TAL for the M 4.4 earthquake on April 25, 1997, at 1.8 and 2.4 km epicentral distances (see Fig. 6.23). Tick marks on vertical axis are 1 cm for GOV and KOY, and 4 cm for TAL. Occasional correlation between at least two well level signals is also obvious. Transient diurnal fluctuations during early March and mid-May 1997 at TAL and KOY probably result from variable loading of water masses in the Koyna River, due to regulatory operations at the Koyna Dam Reservoir

crustal origin for helium. Fissures are inferred to be indicative of the faulting associated with the earthquakes occurring on the Donachiwada Fault. It is also interesting to note that the helium concentrations fall to the background levels of 0.2 ppm within 40–60 m distances from the observed fissures.

To ascertain the dip of the sub-surface fault plane, two boreholes KOY-1 and KOY-2 (Fig. 6.25) located along XX′, were drilled close to traverse DD′. These locations were 45 m and 22.5 m WNW from the location where the highest helium anomaly was found. Both the bore holes were inclined at 43° towards ESE (Fig. 6.25). Steeply inclined slickensides seen in core samples with sub-horizontal slickenlines observed confirm the strike-slip nature of the fault. Reconstruction of the shattered zone with the two boreholes on either side of it indicates that the fault dips WNW at ~ 60° (Fig. 6.26) providing a solution to a long-debated question about the direction of the faulting. The helium studies were carried out in 1995, almost 27 years after the 10 December 1967 earthquake. The observed helium anomalies were very helpful in mapping the rupture caused by the 10 December 1967 earthquake. These field investigations confirmed that the 10 December 1967 Koyna earthquake generated ruptures that reached the surface. This became the 12 SCR earthquake globally, to have documented surface rupturing (Gupta et al. 1999).

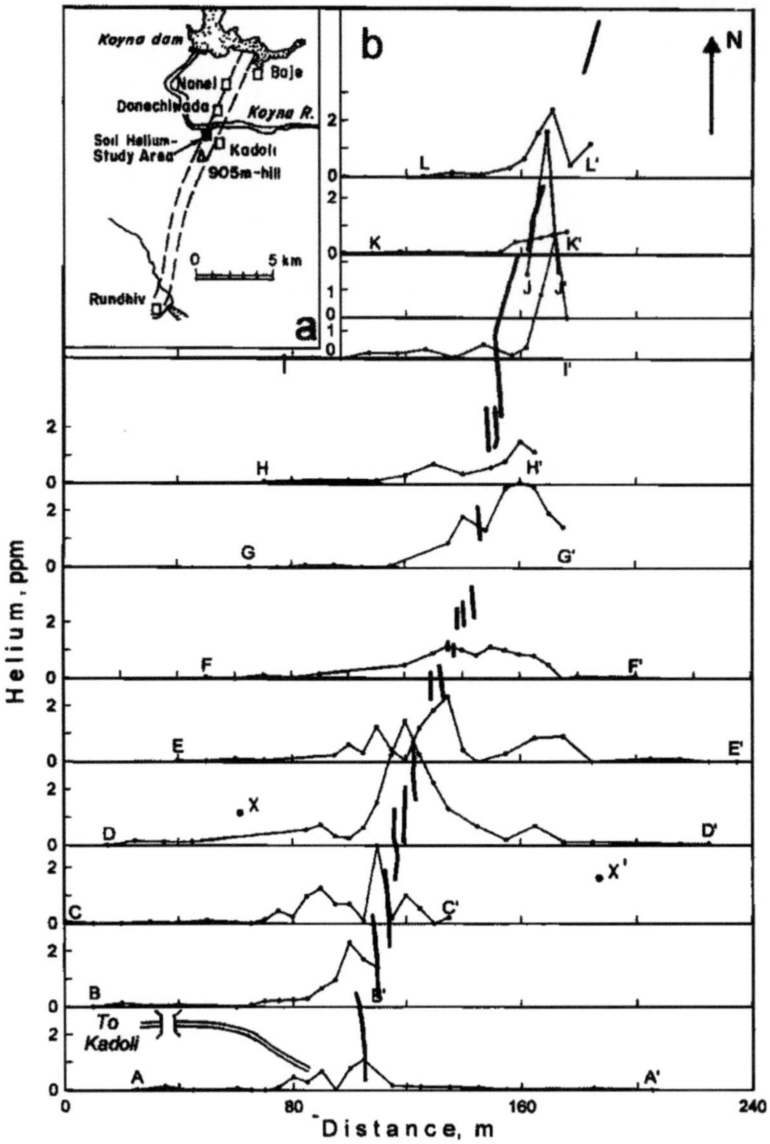

Fig. 6.25 **a** Map of the Donachiwada fault zone (dashed lines) SE of Koyna dam after Talwani (1997) showing the location of the area where soil-helium surveys were conducted (solid rectangle). **b** Coincidence of the helium anomalies with coseismic en-echelon fissures from the 1967 earthquake (thick lines) near Kadoli in the Donachiwada fault zone. Anomalies were established by soil-helium measurements along 12 traverses AA′ to LL′ across the fissure system. The peak values on traverses DD, II and JJ are 4.5, 4.5 and 6.7 ppm respectively. XX′ is the WNW–ESE traverse on which the two boreholes were drilled (Fig. 6.25). The distances between the traverses are on the same scale as for individual traverses

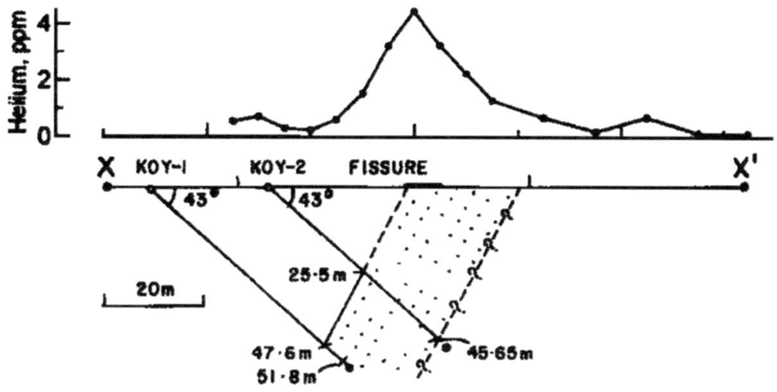

Fig. 6.26 Koyna seismic fault reconstructed by drilling. The two boreholes drilled are on a WNW–ESE traverse perpendicular to the general strike of the fissure zone (shown as XX′ in Fig. 6.25). The helium profile on DD′ is also shown. The bottom of the boreholes is shown by solid circles (KOY-I: 53.3 m; KOY-2: 47.25 m). The brecciated zone was met with in the subsurface between 47.6 and 51.8 m along KOY-1. In KOY-2, it was met with between 25.5 and 45.65 m along the hole. A 60° dip towards WNW for the fault is obtained by connecting the points of first intersection of the breccia zone in the two boreholes and its extrapolation to the surface fissure. The apparent large thickness of the fault zone is guided by the KOY-2 data

References

Bhaskar Rao V, Murty BVS, Murty AVSS (1969) Some geological and geophysical aspects of the Koyna (India) earthquake, December 1967. Tectonophysics 7:265–271

Brahmam NK, Negi JG (1973) Rift valleys beneath Deccan Traps (India). Geophys Res Bull (India) 11:207–237

Chadha RK, Gupta HK, Kumpel HJ, Mandal P, Nageswara Rao A, Kumar N, Radhakrishna I, Rastogi BK, Raju IP, Sarma CSP, Satyamurthy C, Satyanarayana HVS (1997) Delineation of active faults, nucleation process and pore pressure measurements at Koyna (India). Pure Appl Geophys 150:551–562

Committee of Experts (1968) Report on the Koyna earthquake of December 11, 1967—1 and 2. Government of India Press, New Delhi, p 75

Das SR, Ray AK (1972) A note of the photogeological study of Koyna region and part of the west coast in Satara, Ratnagiri and Kolaba districts, Maharashtra. Indian Miner 26:76–79

De Silter LU (1965) Structural geology. In: Series in the geological sciences. McGraw-Hill, New York, 552 pp

Geological Survey of India (1968) A geological report on the Koyna earthquake of 11th December 1967. Calcutta, 242 pp

Grecksch G, Roth F, Kumpel HJ (1999) Coseismic well level changes due to the 1992 Roermond earthquake compared with static deformation of half space solutions. Geophys J Int 138:470–478

Guha SK, Gosavi PD, Varma MM, Agarwal SE, Padale JG, Marwadi SC (1968) Recent seismic disturbances in the Koyna Hydroelectric Project, Maharashtra, India, 1. Report. C.W.P.R.S., 16 pp

Guha SK, Gosavi ED, Varma MM, Agarwal SE, Fadale JG, Marwadi SC (1970) Recent seismic disturbances in the Shivajisagar Lake area of the Koyna Hydroelectric Project, Maharashtra, India, 2. Report. C.W.P.R.S., 25 pp

Guha SK, Gosavi ED, Nand K, Eadale JG, Marwadi SC (1974a) Koyna earthquakes (Oct. 1963 to Dec. 1973). Central Water and Power Research Station, Khadakwasla, Poona, 24 pp

Guha SK, Gosavi ED, Agarwal BNE, Eadale JG, Marwadi SC (1974b) Case histories of some artificial crustal disturbances. Eng Geol 8:59–77

Gupta HK (1983) Induced seismicity hazard mitigation through water level manipulation: a suggestion. Bull Seismol Soc Am 73:679–682

Gupta HK (1992) Reservoir induced earthquakes. Elsevier Scientific Publishing Company, Amsterdam, p 364

Gupta HK (2002) A review of recent studies of triggered earthquakes by artificial water reservoirs with special emphasis on earthquakes in Koyna, India. Earth Sci Rev 58:279–310

Gupta HK, Rastogi BK (1976) Dams and earthquakes. Elsevier Scientific Publishing Company, Amsterdam, 229 pp. Translated into Russian in 1979 and Chinese in 1980

Gupta HK, Combs J(1976) Continued seismic activity at the Koyna reservoir site. India Eng Geol 10:307–313

Gupta HK, Narain H, Rastogi BK, Mohan I (1969) A study of the Koyna earthquake of December 10, 1967. Bull Seismol Soc Am 59:1149–1162

Gupta HK, Rastogi BK, Narain H (1971) The Koyna earthquake of December 10, 1967: a multiple seismic event. Bull Seismol Soc Am 61:167–176

Gupta HK, Rao CVRK, Rastogi BK (1980) An investigation of earthquakes in Koyna Region, Maharashtra, for the period October 1973 through December 1976. Bull Seismol Soc Am 70(5):1833–1847

Gupta HK, Rao RUM, Srinivasan R, Rao GV, Reddy GK, Dwivedi KK, Banerjee DC, Mohanty R, Satyasaradhi YR (1999) Anatomy of surface rupture zones of two stable continental region earthquakes, 1967 Koyna and 1993, Latur, India. Geophys Res Lett 26(13):1985–1988

Gupta HK, Radhakrishna I, Chadha RK, Kumpel HJ, Grecksch G (2000) Pore pressure studies initiated in area of reservoir-induced earthquakes in India. EOS Trans 81(14):145, 151

Harpster RE, Cluff LC, Lovegreen JR (1979) Active faulting in the Deccan plateau near Koyna Nagar, India. Geol Soc Am Abstr Progr 11(7):438–439

Kailasam LN, Murthy BGK (1969) Geophysical investigations in earthquake affected areas of Koyna, Satara district, Maharashtra. Mem Geol Surv India 100:117–122

Kailasam LN, Pant PR, Lahiri SM (1969) Seismic investigations in the earthquake affected areas of Koyna and neighborhoods, Satara district, Maharashtra. Mem Geol Surv India 100:123–126

Kailasam LN, Murthy BGK, Chayanulu AYSR (1972) Regional gravity studies of the Deccan Trap areas of Peninsular India. Curr Sci 41(11):403–407

Kaiser J (1953) Erkenntnisse and Folgerungen aus der Messung von Gerauschen bei Zugbeanspruchung von metallischen Werkstoffen. Archiv Fur Das Eisenhuttenwesen 24:43–45 (in German)

Krishna J, Chandrasekharan AR, Saini SS (1969) Analysis of Koyna accelerograms of December 11, 1967. Bull Seismol Soc Am 59:1719–1731

Krishnan MS (1960) Geology of India and Burma. Higginbothams, Madras, 604 pp

Langston CA (1976) A body wave inversion of the Koyna, India, earthquake of December 10, 1967, and some implications for body wave focal mechanisms. J Geophys Res 81(14):2517–2529

Mane PM (1967) Earth tremors in the Koyna Project area. In: 9th international congress on large dams, Istanbul, Commission Internationale des Grands Barrages, Paris, pp 509–518

Narain H, Gupta HK (1968) The Koyna earthquake. Nature 217:1138–1139

Rastogi BK, Talwani P (1980) Spatial and temporal variations in t_s/t_p at Monticello Reservoir, South Carolina. Geophys Res Lett 7:781–784

Roeloffs EA (1988) Fault stability changes induced beneath a reservoir with cyclic variations in water level. J Geophys Res 93(B3):2107–2124

Simpson DW, Negmatullaev SK (1978) Induced seismicity studies in Soviet Central Asia. U.S. Geol Surv Earthq Inf Bull 10(6):208–213

Simpson DW, Negmatullaev SK (1981) Induced seismicity at Nurek Reservoir, Tadjikistan, USSR, Bull Seismol Soc Am 71(5):1561–1586

Singh DD, Rastogi BK, Gupta HK (1975) Surface-wave radiation pattern and source parameters of the Koyna earthquake of December 10, 1967. Bull Seismol Soc Am 65:711–731

Sykes LR (1970) Seismicity of the Indian ocean and a possible nascent island arc between Ceylon and Australia. J Geophys Res 75:5041–5055

Talwani P, Kumar Swamy SV, Sawalwade CB(1996). The revlauation of seismicity data in the Koyna-Warna area, 1963–1995, Technical Report, University of South Carolina, p 343

Talwani P (1997) Seismotectonics of the Koyna-Warna area, India. Pure Appl Geophys 150:511–550

Thirlaway HIS (1963) Earthquake or explosion. New Sci 18:311–315

Wyss M, Brune JN (1967) The Alaska earthquake of 28, March 1964: a complex multiple rupture. Bull Seismol Soc Am 57:1017–1023

Yoshikawa S, Mogi K (1981) A new method for estimation of the crustal stress from cored rock samples: Laboratory study in the case of uniaxial compression. Tectonophysics, 74, 323–339

Studies in the Koyna, India, 2004–2024

7

This chapter deals with the work having been carried out in the Koyna region during the period 2004–2024. The CSIR-National Geophysical Research Institute, Hyderabad, India set up a network of 7 three component seismic stations in the vicinity of RTS area of Koyna and the seismic data were communicated to Hyderabad in near real time. M ~ 4 earthquakes were found to be preceded by precursory swarms. Recognition of swarm preceding the mainshock in real time led to making successful earthquake forecasts of M ~ 4 earthquakes. Globally, knowledge about the physical properties of rocks and fluids in the fault zone before, during and after the occurrence of an earthquake are lacking. Koyna has been found to be a suitable locale for such investigations as the earthquakes are confined to a region of 20 km × 30 km, earthquakes occur at a shallow depth (mostly between 2 and 9 km), the crust is cool (~ 130 °C at 6 km depth), there is no other source of earthquakes within 50 km of Koyna Dam, the region is totally accessible for all kind of measurements. With this in view, a detailed proposal was made to the International Continental Drilling Program (ICDP) in 2010. It was accepted and a dedicated workshop at CSIR-NGRI, Hyderabad and field visit to Koyna were held in 2011. Attended by 74 experts, the workshop endorsed the proposal and suggested improving earthquake locations, carrying out airborne gravity and magnetic surveys, LiDAR and magnetotelluric surveys. All the suggested work was carried out over the next three years and reviewed in the next workshop held at Koyna in 2014. Satisfied with the progress of the work, the 2nd ICDP workshop gave a go ahead for putting a 3 km deep Pilot Borehole at an appropriate location. Based on earthquake distribution, logistic requirements and vicinity to the Donachiwada Fault that had hosted the 10 December 1967 M 6.3 and several M ~ 5 earthquakes, a location was chosen for the Pilot Borehole. The borehole was spudded in December 2016 and completed in June 2017. Several measurements were made during drilling and core samples from critical zones were collected for further studies. A Post-Operational ICDP workshop was held at Pune, India during 14–16 October 2017. It was attended by 67 participants including 19 from Canada, France, Germany, Italy, Japan, the Czech Republic, South Korea, Spain, Switzerland, and the USA. There was a total satisfaction with the results obtained and plans of further work leading to designing the proposed 7 km deep Main Borehole were discussed. Participants of the workshop visited the Borehole Geophysics Research Laboratory at Karad. The Core Repository was of immense interest. On the final day, detailed recommendations for the future course of action were made focusing on developing of Koyna RTS information system; characterization of Donachiwada fault; deployment of a vertical string of seismometers below 1500 m in the Pilot Borehole; a variety of hydrogeological/geophysical/geological measurements to be made; detailed study of the core samples; setting up of a reference borehole outside Koyna RTS area; and preparation of a detailed plan of the proposed deep borehole laboratory at a depth of ~ 7 km. Overall, the placing of the borehole at a very appropriate location and carrying out all the necessary measurements during the drilling of 3-km-deep borehole were appreciated. In the year 2017, the completion of 5 decades of RTS at Koyna, India, was marked with commemoration of the 10 December 1967 earthquake in the session on Anthropogenic Seismicity at Kobe, Japan during the joint assembly of the International Association of Seismology and Physics of the Earth's Interior (IASPEI) and the International Association of Geodesy (IAG) during 30 July–4 August 2017. The Anthropogenic Seismicity was the second most attended session at Kobe with over a dozen presentations dealing with RTS at Koyna, India. The Journal of Geological Society of India dedicated December 2017 issue with 25 articles dealing with different aspects of RTS at Koyna, India. A proposal for a session was made to the 2017 annual fall meeting of the American Geophysical Union, at New Orleans, Louisiana, entitled "Fifty Years since the World's Largest Reservoir Triggered Seismic Event, the 10 December 1967 M 6.3 Koyna

H. K. Gupta, *Artificial Water Reservoir Triggered Seismicity*, https://doi.org/10.1007/978-3-032-05873-7_7

Table 7.1 Events preceding M ≥ 4.0 earthquakes in Koyna region (taken from Gupta 2021)

Main earthquake	Duration of nucleation period (h)	No. of events before the main earthquake				
		M 1.0–1.9	M 2.0–2.9	M 3.0–3.9	Largest earthquake	50 h prior to beginning of nucleation period
30 August 2005 M 4.8	110	13	8	0	2.6	2
13 November 2005 M 4.0	160	18	8	1	3.0	3
26 December 2005 M 4.2	150	22	8	0	2.9	6
17 April 2006 M 4.7	380	40	10	0	2.7	2

Earthquake—Lessons Learnt". It was merged with the session on "Induced Seismicity in the United States and Canada". The proceedings of this session were published as a Special Issue of the Bulletin Seismological Society of America in November 2018 with the heading, "Fifty Years after the 1967 Koyna Earthquake—Lessons Learned about Reservoir Triggered Seismicity (RTS)" (Rao et al. 2018). The work was going on in full swing after the third ICDP workshop when the COVID-19 pandemic put brakes on the progress and broke the momentum. The 3 years of uncertainty (2019–2022) has resulted in certain setbacks in meeting the planned targets. Some of the planned targets and the timelines were worked out keeping in view the conditions prevailing in the country and in the world. I am hopeful, that a dream of drilling a 7 km deep borehole right into the heart of the seismic zone to catch a *Live* earthquake may become a reality someday.

7.1 Earthquake Forecasts

A detailed description of earthquake forecasts at Koyna has been provided by Gupta (2021). From among all the precursors investigated globally, observation of the foreshocks is one of the most important precursors and has a potential of making forecasts, as highlighted by Dodge et al. (1996) and Helmestetter et al. (2003). It has been observed by Ohnaka (1992) that the immediate foreshock activity constitutes the nucleation process leading to the occurrence of the main shock. Way back in 2001, Gupta (2001) had postulated that earthquake forecast may be feasible at Koyna, India. The seismic activity in the Koyna region has been monitored in real time since 2005 with a network of seismic stations operating in the immediate vicinity of the reservoir (Gupta et al. 2007). The network consists of 7 three components seismic stations, 4 of them being broad band (CMG3ESP) and the remaining three short-period (L4-3D) seismometers, equipped with REFTEK recorders and time synchronized by GPS. This has permitted location of epicenters with an accuracy of ± 500 m. From August 2005 data were communicated to CSIR-NGRI and analyzed daily. To examine whether there are recognizable foreshocks preceding $M \geq 4$ earthquakes in the Koyna region, a detailed study was undertaken. It was observed that $M \sim 4$ earthquakes in the Koyna region are generally preceded by a well-defined foreshock sequence. Table 7.1 demonstrates the learning phase where four earthquakes of *M* 4.0–4.8 were preceded by a significant foreshock activity confined to a tight cluster within 10 km radius (Fig. 7.1a–d). This led to the confidence that if the formation of the nucleation is identified in real time, it could lead to a short-term forecast. Such a situation arose in middle of May 2006. Earthquakes during 11–16 May 2006 are depicted in Fig. 7.2. A total of some 50 events of $M \geq 1.0$, with the largest event being of M 2.7, with shallow focal depths (2–8 km) were found clustered in a small area in the preceding 107 h. This led to the inference that the region was experiencing a nucleation phase. The following forecast was made:

> On the basis of the data available from seven seismic stations operating in the Koyna region, we have identified a nucleation which started on 12 May 2006. This may lead to the occurrence of an M ~ 4 earthquake in the next 15 days. This shallow earthquake (focal depth less than 8 km) will occur within a radius of 10 km centered at 17.1° N, 73.8° E. On the basis of our previous experience of studying nucleation-preceding earthquakes in the Koyna region, we expect this earthquake to occur over the next 15 days-time (till 31 May 2006), with a 50% probability.

It was communicated to the Editor of Current Science and the Secretary of the Ministry of Earth Sciences, Government of India. To avoid panic, no information was given to the public. An M 4.2 earthquake did occur on 21 May 2006 well within the forecasted parameters (Table 7.2, Gupta et al. (2007)). After making successful forecast of M ~ 4 earthquake in Koyna during 2006, the effort continued and a total of six forecasts were successfully made, and the magnitude of these forecasted earthquakes ranged from M 4 to 5. In a later study, Gupta et al. (2011) noted that during the period August 2005 through May 2010, a total of 29 earthquakes in the magnitude range of 3.5–5.1 occurred in the Koyna region, which could broadly be placed in three zones (Fig. 7.3). While Zone A has been most active during this period accounting for 18 of these

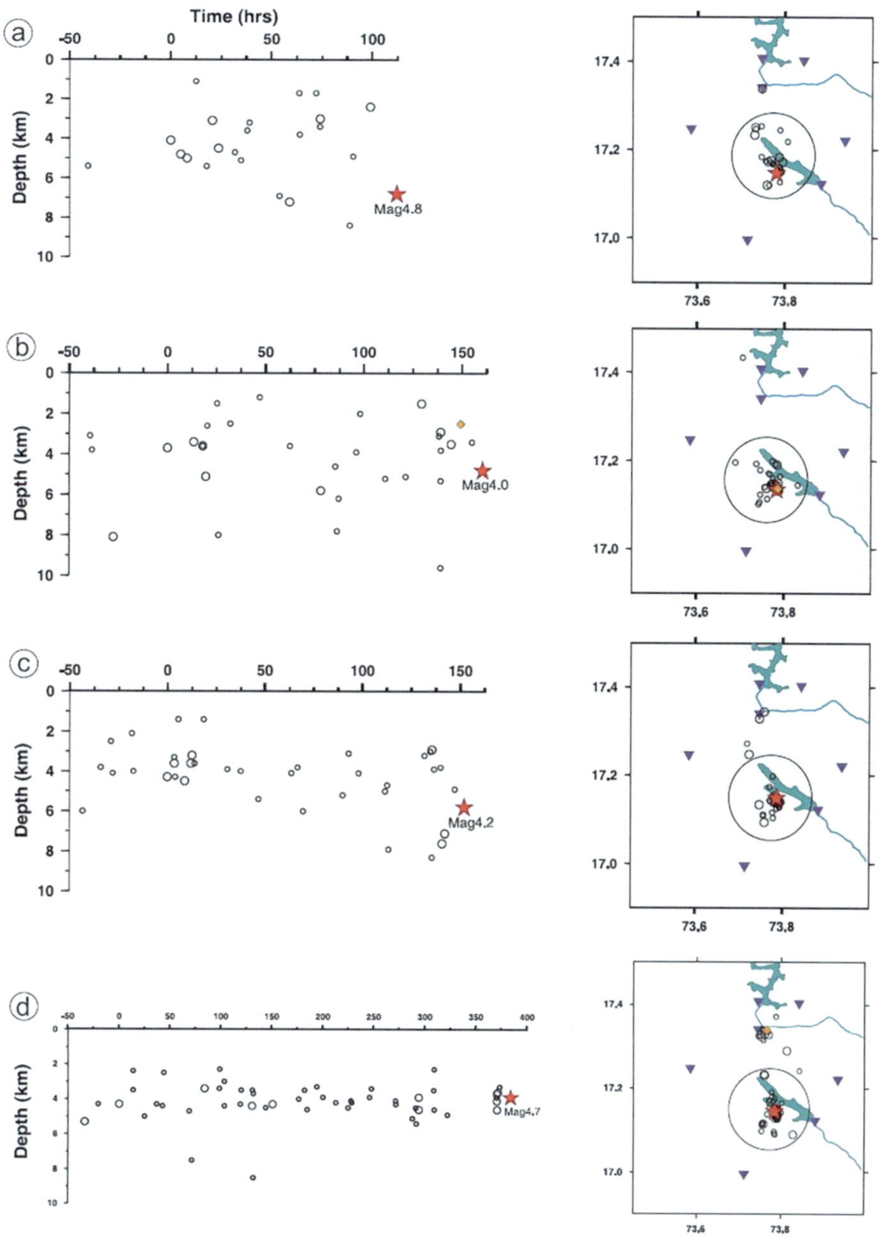

Fig. 7.1 Seismic activity for the period 50 h before the start of nucleation till the occurrence of M $\geq$ 4 earthquakes in Koyna region during August 2005–April 2006. Foreshocks clustering before the mainshocks in a region of 10 km radius (right). Temporal distribution of events within the circle with the depth. ○, O, ◆ and ★ are events in M 1.0–1.9, 2.0–2.9, and 3.0–3.9 ranges and main earthquake, respectively (left). The onset of identified nucleation period is given "0" time mark. Nucleation period lasted for 110, 160, 150, and 380 h for **a** 30 August, **b** 13 November, and **c** 26 December 2005 and **d** 17 April 2006 earthquakes, respectively (Table 7.1)

31 earthquakes, Zones B and C had 5 and 6 earthquakes in the 3.5–5.1 magnitude range. Table 7.3 provides information as to which all these 31 earthquakes were preceded by nucleation. While earthquakes in Zone A are found to be mostly preceded by nucleation, the same is not the case with Zones B and C. It must also be mentioned that all M ~ 4 earthquakes were not forecasted and there was no false forecast. This amply supports the complexities in earthquake forecasting. Even in a small region, like Koyna, there are pockets of rock volumes where precursory nucleation occurs, but that is not true for the entire Koyna region (Gupta 2021) (Table 7.3).

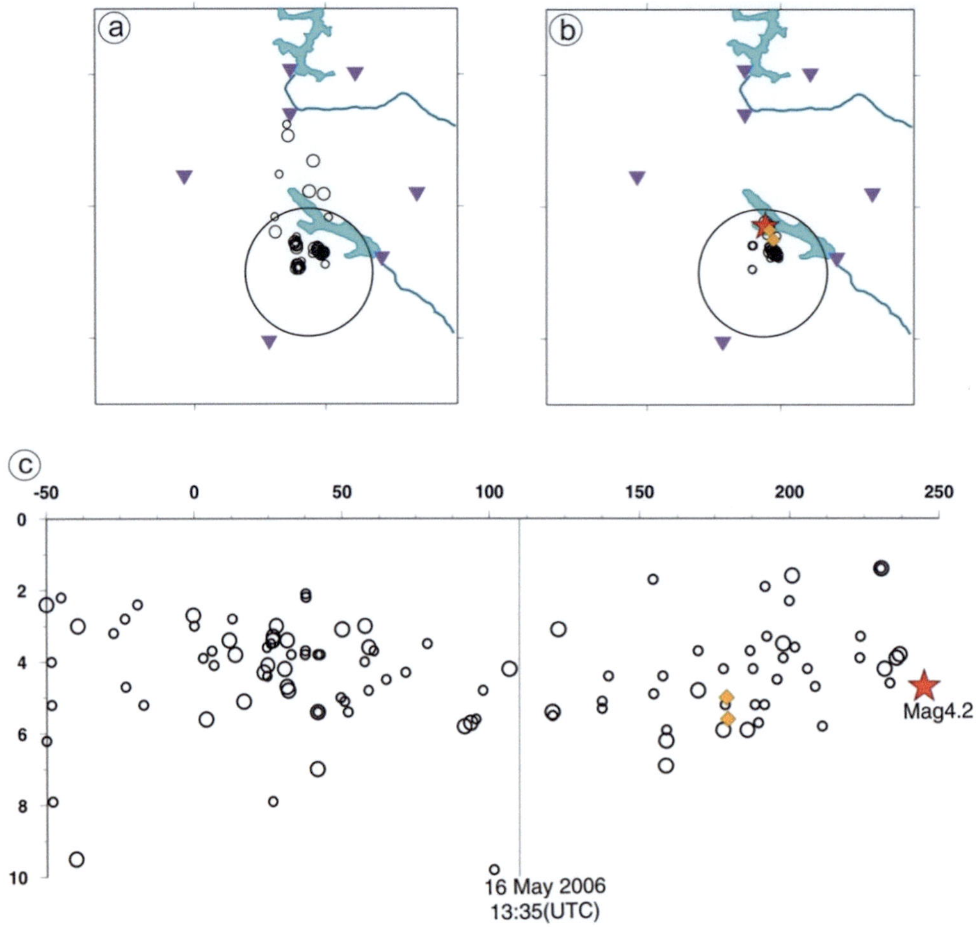

Fig. 7.2 **a** Events that occurred during 107 h (11–16 May) preceding identification of nucleation. **c** Temporal-depth plot of the events within the 10-km-radius circle. Nucleation was inferred to have started 107 h before, where 0 h is put on the time axis. **b** Seismic activity from 13:35 UTC on 16 May till the occurrence of the M 4.2 earthquake on 21 May 2006, on the right side of **c**. ○, O, ♦ and ★ are events in M 1.0–1.9, 2.0–2.9, and 3.0–3.9 ranges and main earthquake, respectively

Table 7.2 Forecast of 21 May 2006 M 4.2 earthquake in Koyna on 16 May 2006

	Forecast parameters	Occurrence
Epicenter	Within 10 km radius of 17.1° N, 73.8° E	17.171° N, 73.777° E
Magnitude	~ 4	4.2
Time	Within 15 days of 16 May 2006, i.e. until 31 May 2006	20:29:01 UTC On 21 May 2006
Focal depth	Less than 8 km	4.7 km

7.2 Studies to Elucidate Genesis of RTS[1]

A major requirement to comprehend the genesis of RTS is near-source studies. As a matter of fact, globally, there is a need of such studies to validate earthquake models.

The correspondence between water levels in the reservoir, rate of loading, highest water levels reached and other water level parameters and RTS has been established at several sites globally (Gupta 1992, 2002, 2021; Simpson and Narasimhan 1990; Yadav et al. 2017 and many others). However, the triggering mechanism is still not well understood. It is important to investigate the physical properties of rocks and fluids in the fault zone through near-field studies. Koyna, India, appeared as an ideal site for undertaking such studies (Gupta and Nayak 2011). The reasons include continued RTS at Koyna since the impoundment of Koyna Dam in 1962; the seismic activity being confined to a small area of 30 × 20 km; there being no other source

[1] Based on Gupta (2022).

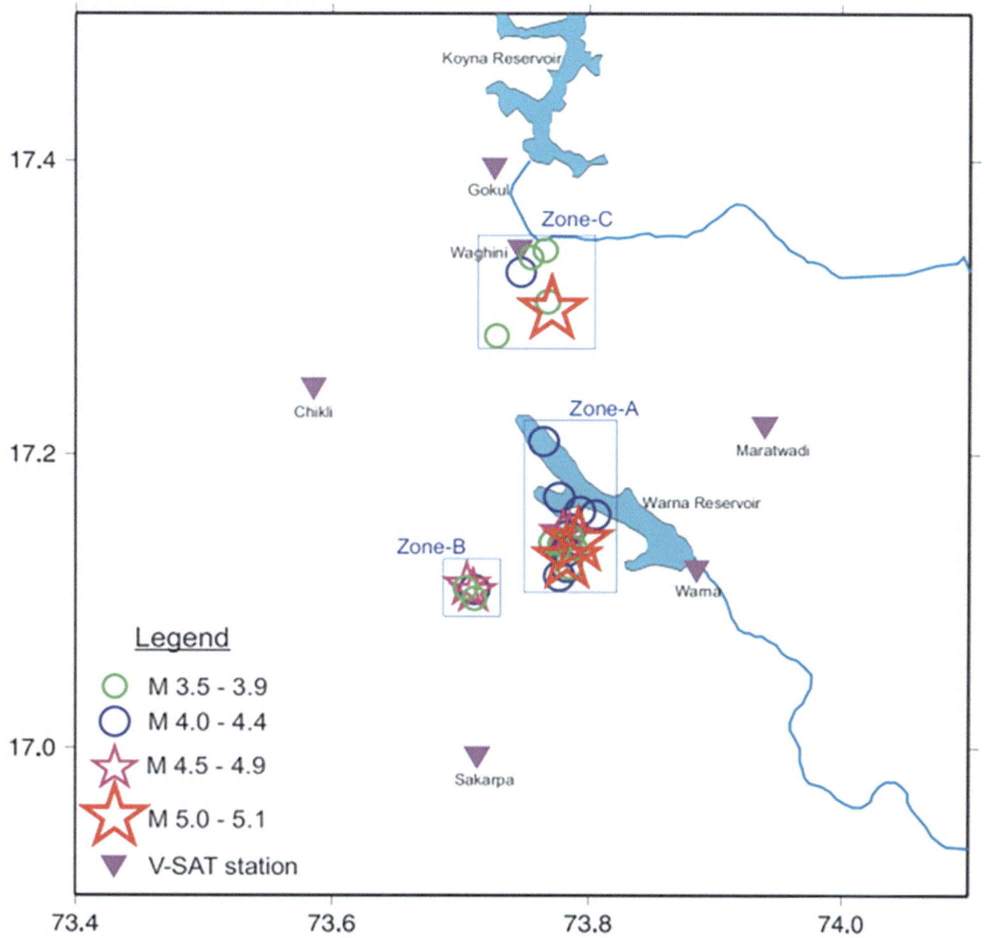

Fig. 7.3 Earthquakes of magnitude range 3.5–5.1 in the Koyna-Warna region during August 2005–May 2010. Zone 'A' indicates area in which all of the M 3.5 and above earthquakes preceded by nucleation, whereas in Zone 'B' and Zone 'C' nucleation was not observed (after Gupta et al. 2011)

of earthquakes within 100 km of Koyna Dam; the earthquakes mostly occurring at shallow depths of 2–9 km; the area being accessible to all kinds of observations and the measurements; and a relative cool upper crust with estimated temperatures being limited to ~ 130 °C at a depth of 6 km, permitting long time in situ measurements using standard electronic equipment. A proposal was submitted to International Continental Drilling Program (ICDP) for holding a workshop to discuss the suitability of Koyna for setting up of a Borehole Laboratory at Koyna for near-field studies in 2010. After the approval of this proposal, a workshop was organized at Hyderabad and Koyna during 21–26 March 2011. The following were the objectives of this workshop (Gupta 2018):

1. To exchange lessons learned from investigations on RTS worldwide, including Koyna, at an international forum.
2. To fine-tune the scientific objectives for deep drilling at Koyna, a classical RTS site in an intra-plate setting.
3. To prepare a complete drilling plan.
4. To plan the entire array of measurements/monitoring opportunities provided by deep drilling in consultation with national and international experts.
5. To develop a full proposal on scientific drilling at Koyna.

The workshop was attended by 76 participants including 26 from outside India, who have had experience of working in related areas, globally. It was agreed, with unanimity, that Koyna, India, is the most suitable site for undertaking near-field studies through setting up of bore-hole laboratory for comprehending RTS. The suggestions made for additional work to be carried out before drilling of the borehole laboratory included:

(1) improving the location of the hypocenters in the region by adding more seismic stations,

Table 7.3 The earthquakes in chronological order during August 2005–May 2010 that occurred in Zone 'A', Fig. 7.3 (after Gupta et al. 2011)

S. No.	Day of occurrence	Origin time (hours, minutes, seconds)	Latitude (° N)	Longitude (° E)	Depth (km)	Magnitude (M_L)	Nucleation
1	14-08-2005	06 33 28.0	17.161	73.793	5.0	4.2	Preceded
2	30-08-2005	08 53 18.1	17.147	73.780	6.8	4.8	Preceded and forecasted
3	13-11-2005	03 56 08.0	17.134	73.783	4.8	4.0	Common nucleation, being only one week apart
4	20-11-2005	18 50 42.0	17.209	73.765	5.9	4.1	
5	26-12-2005	10 46 06.3	17.149	73.786	5.8	4.2	Preceded
6	17-04-2006	16 39 59.41	17.145	73.785	3.9	4.7	Preceded
7	21-05-2006	20 29 01.2	17.171	73.777	4.7	4.2	Preceded and forecasted
8	19-01-2007	19 59 23.39	17.143	73.786	2.5	3.8	Preceded
9	20-08-2007	19 15 54.5	17.145	73.784	3.6	4.3	Preceded
10	20-08-2007	19 31 45.4	17.136	73.789	4.3	3.6	Aftershock of # 9
11	14-09-2007	01 05 21.2	17.139	73.778	4.4	3.5	Preceded
12	14-11-2009	13 03 33.8	17.109	73.788	9.6	5.0	Preceded
13	14-11-2009	13 34 34.8	17.115	73.777	1.7	4.0	Aftershock of # 12
14	12-12-2009	11 51 24.8	17.126	73.788	5.5	5.1	Preceded and forecasted
15	12-12-2009	21 55 3.2	17.146	73.770	1.3	4.0	Aftershock of # 14
16	16-12-2009	05 27 2.7	17.140	73.771	3.5	3.5	Aftershock of # 14
17	23-12-2009	03 49 28.5	17.123	73.783	3.2	4.0	Aftershock of # 14
18	12-01-2010	09 25 29.2	17.123	73.785	3.2	3.8	Preceded

(2) conducting airborne gravity and magnetic surveys and surface seismic and magneto-telluric surveys for a better understanding of local subsurface structure, and
(3) LiDAR coverage of the area to get bare-earth surveys and decipher surface lineaments/faults as well as drainage patterns. It was further suggested that after these investigations have been carried out, the new knowledge gained would be helpful in developing the plan of setting up a bore-hole laboratory.

During the next three years (2011–2014), the suggested investigations had been carried out, as depicted in Fig. 7.4. These included (Gupta 2022):

1. Nine exploratory boreholes were drilled through the entire basalt column and penetrating to 300–500 m into the granitic basement extending to a depth of ~ 1500 m. Eight of them were instrumented with three component seismometers placed at the bottom of the boreholes.
2. The entire RTS zone and nearby area were covered by detailed magneto-telluric surveys for estimating the thickness of the basalt cover and its lateral variations.
3. For the first time, airborne gravity gradient surveys were carried out over the Koyna region. Additional, airborne magnetic surveys were also conducted for a better understanding of the subsurface structure.
4. The region was covered by LiDAR surveys which provided a bare-earth visualization, and mapping of fault-related geomorphology.
5. Out of the 9 drilled boreholes, 6 were instrumented with matched 3-component seismometers for improving hypocentral parameters. All the seismometers were placed at around 1500 m depth, well below the basalt cover.
6. Detailed heat flow measurements were carried out in the boreholes to have a better estimate of subsurface temperatures.

The above-mentioned investigations were extremely helpful in our understanding the subsurface structure as well as improving the location of earthquakes to ± 300 m accuracy. In the following, salient features are provided.

7.3 Improving the Location of Earthquakes

Since 2005, the CSIR-National Geophysical Research Institute (NGRI) had been operating several seismic stations in and around the Koyna earthquake zone. Their number varied and was up to 20. The volcanic cover prevailing all over the Koyna

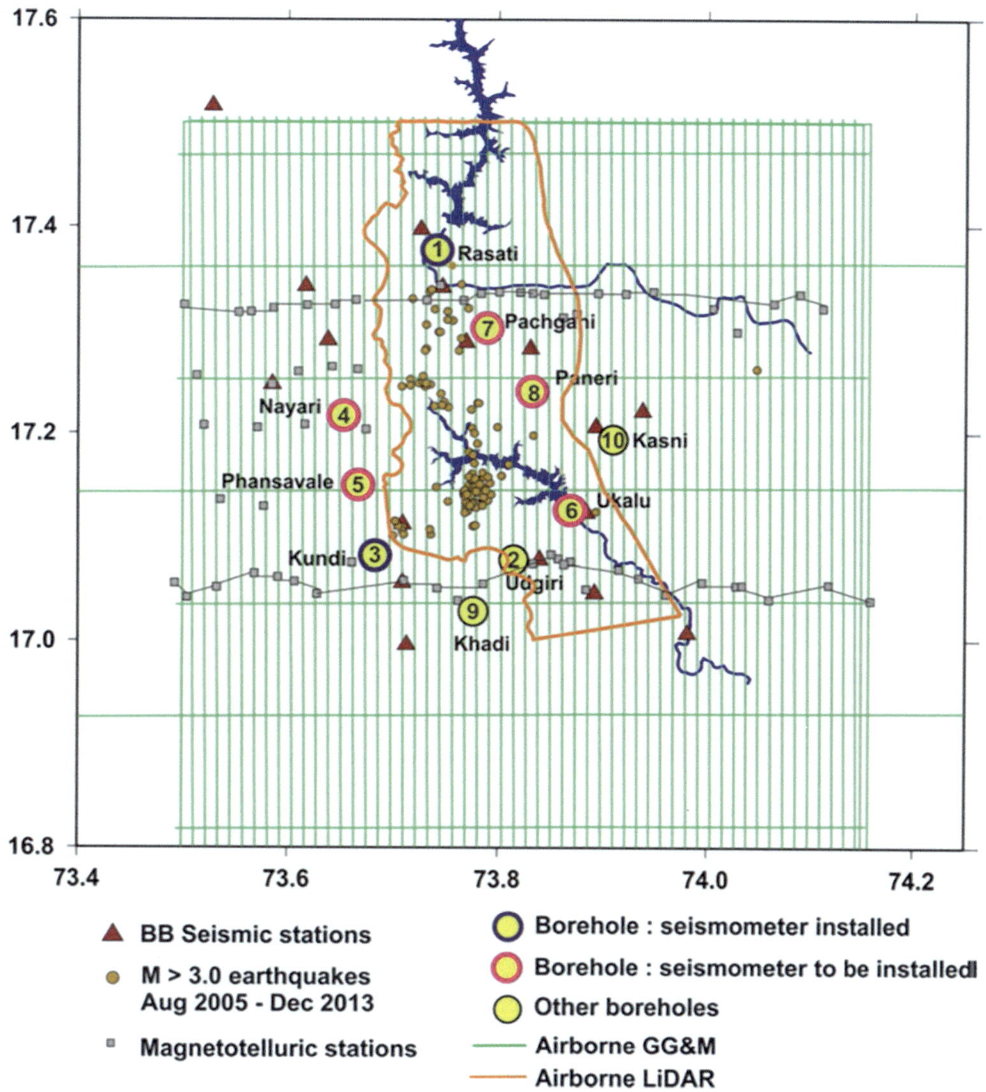

Fig. 7.4 The study area indicating the installation or deployment plan of the various experiments undertaken in the preparatory phase of the deep drilling program. Earthquakes of M $\geq$ 3.0 occurred during August 2005–December 2013. MT-1 and MT-2 indicate the magnetotelluric profiles passing through Rasati in the north and Udgiri in the south respectively

earthquake zone adds noise to the seismic signals and location of the epicenter of the Koyna earthquakes was not better than ~ 800 m. To improve the location, 9 boreholes were drilled reaching ~ 400 m below the basalt cover (Fig. 7.4). Five of these bore holes were instrumented with matched 3-component seismometers during March–June 2015 and three additional boreholes were instrumented in later years. Five borehole seismometers were installed in the seismically active Koyna region at depths varying from 1134.0 to 1522.5 m. The seismometers, with a natural frequency of 4.5 Hz are connected to REFTEK digitizers with time synchronization through Global Positioning System. Recording is being done in a continuous mode with a sampling rate of 250 samples per second (Shashidhar et al. 2014). Consequent to setting up of the borehole seismic stations, the epicenter location has improved to within 200 m. All earthquakes of M $\geq$ 0.5 are being located since the commissioning of this borehole network of seismic stations (Shashidhar et al. 2016).

7.4 Airborne Gravity Gradient and Magnetic Surveys[2]

Airborne gravity-gradient and magnetic (AGGM) surveys provide measurements from which density structures and magnetic anomalies can be delineated. Using such surveys, we unraveled the sub-surface structure in the Koyna–Warna

[2] Based on Elliott et al. (2020).

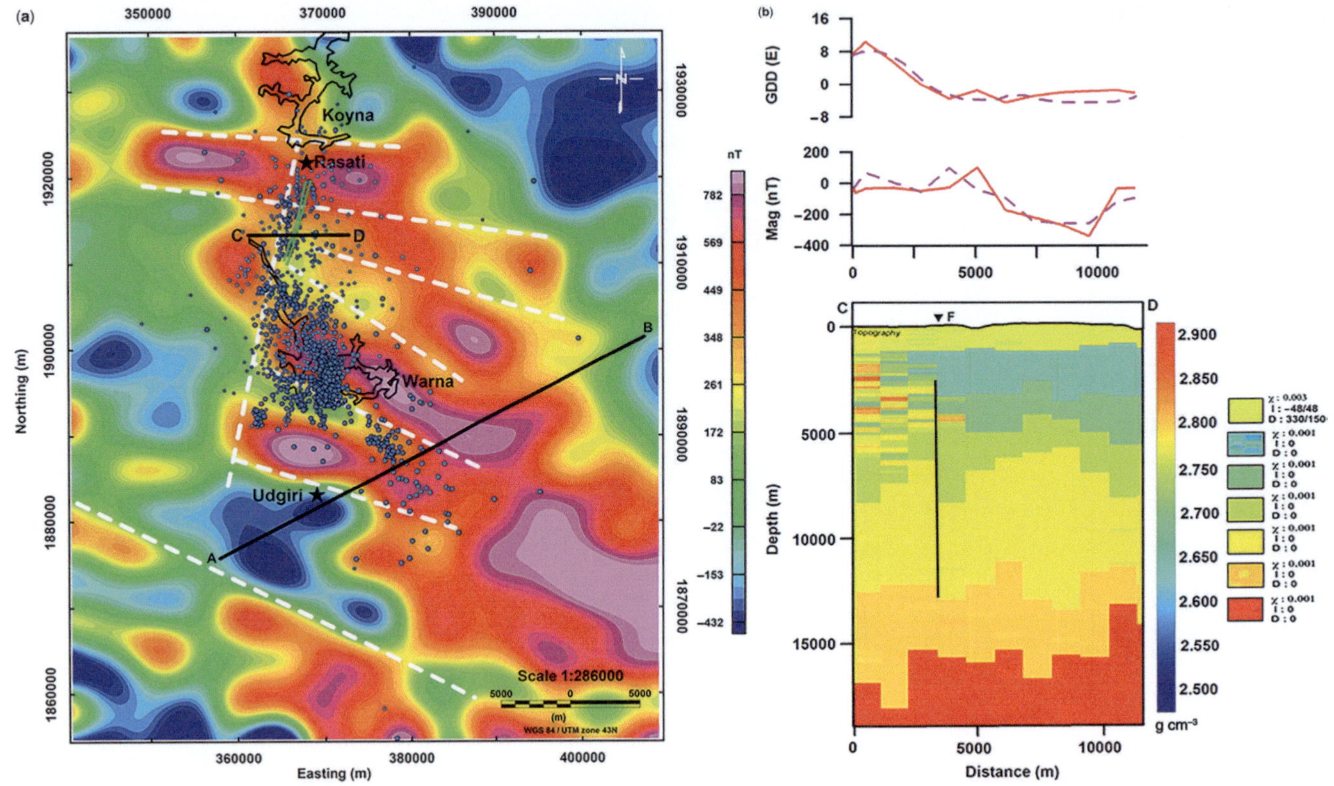

Fig. 7.5 **a** Low pass (0.10 km) filtered magnetic anomaly map of the region; earthquakes of M $\geq$ 2.0 between 2005 and 2015 are plotted (blue dots). The Koyna and Warna reservoirs and the borehole locations of KBH-1 at Rasati and KBH-2 at Udgiri are shown. AB marks a section across the region highlighting the alignments of the most prominent magnetic anomalies. CD is a section marked across the Donachiwada fault inferred from seismology. **b** Modelled subsurface configuration across the Donachiwada fault (CD marked in **a**) from AGGM data; the top panel shows the observed (dashed) and modeled (continuous) GDD anomaly for this section; the middle panel shows observed (dashed) and modeled (continuous) magnetic anomaly across this section; bottom panel shows the density-susceptibility model, the dashed line indicates the subsurface vertical structure below the Donachiwada fault, which is derived from the contrasts in the model parameters

region to address the potential controls on the pattern and distribution of observed RTS in the area. The region covered is shown in Fig. 7.4 and is a draped surface about 120 m above ground level at an interval of ~ 1 km. The magnetic anomalies are predominately caused by the basaltic layer, which varies in thickness from 400 to 1600 m (Gupta et al. 2016). The magnetic data were filtered with a cut-off wavelength of 10 km to focus on the deeper sources of these anomalies. The resulting map broadly depicts the subsurface structure (Fig. 7.5a, b). It is characterized by a prominent NW–SE trend in the southern region and a NNE–SSW trend in the center, flanked by a prominent negative anomaly. This anomaly is centered close to the Udgiri borehole (Fig. 7.5a). An east–west trend of the anomaly is observed close to the southern end of the Koyna Dam. It has been also found that this east–west trending anomaly disappears when data are filtered with a higher cut-off wavelength leading to the inference of the source of these anomalies being shallower compared to the north–south and northwest–southeast trending anomalies. The northwest-southeast and north–south trending features are prominently identified in the central area of RTS in the filtered anomaly plots. This is consistent with the earthquake cluster in the vicinity of the Western Ghats Escarpment. The subsurface structure below the Donachiwada fault (considered to be the main causative fault of RTS at Koyna) has been inferred using the AGGM data as depicted in Fig. 7.5b. It corresponds to a low in the observed GDD (Vertical Gravity Gradient) and magnetic field, as the topmost layer is basalt with high magnetic susceptibility and density increases with depth. The delineated vertical block structure is consistent with the fault inferred from earthquake data.

7.5 LiDAR Surveys

The Koyna-Warna region is characterized by rugged topography. A thick vegetation cover prevails all over the region. This situation makes observation of surface manifestation of tectonic events difficult to decipher. A lineament map using the LANDSAT imagery was prepared by Langston (1981) which indicated a predominant NNW and NNE oriented en echelon zone following a trend parallel to Western Ghats

Escarpment (WGE) in the Koyna-Warna region. InSAR data has provided surface observations of vertical crustal deformation associated with the filling and emptying cycles of the Koyna and Warna reservoirs. Shuttle Radar Topography Mission (SRTM) data have been used by Das and Mallik (2020) to comprehend morphometric parameters of the Koyna-Warna region to outline areas which have witnessed tectonic movements on the right flank of the Shivaji Sagar Lake formed by the Koyna Dam. High resolution LiDAR (Light Detection and Ranging) has become a very useful tool for topographic data accusation and its use in Earth Sciences (Krishnan et al. 2011). LiDAR data are particularly useful for identifying and tracing surface manifestation of faulting associated with tectonic movements, particularly earthquakes (Zeilke et al. 2012). Although, the Koyna 1967 earthquake generated movements along the Donachiwada Fault's surface manifestation had been earlier traced by observing helium anomalies associated with the rupture reaching the Earth surface (Gupta et al. 1999), it was thought to be useful to conduct LiDAR surveys covering the entire earthquake zone in the Koyna-Warna region to get bare topography of the region and look for the surface ruptures that may be associated with the earthquakes occurring in the Koyna-Warna region. The rugged topography and dense vegetation prevalent in the Koyna-Warna region made it very difficult to trace possible earthquake generated ruptures at the surface.

A detailed LiDAR survey was conducted over 1064 km^2 of area (Fig. 7.6) covering the Koyna-Warna Seismic Zone (KWSZ). A month before the survey the Ground Control Points and the Primary Base survey were established (Arora et al. 2017). A good data quality was ascertained with sufficient GPS lock and quality monitoring of the internal measurement unit. The flight height was on average 650 m above the ground level and the speed was 90 knots/h. Arora et al. (2017) have provided all the details of the survey. A 60% overlap and distance between two photographs as 205 m was maintained. The point density has been 6.09/m^2 on the ground. A final vertical accuracy of 10 cm, horizontal accuracy of 15 and Orthomosaic Ground Sampling Distance of 10 cm was achieved. The Bare Earth Model (BEM) obtained is depicted in Fig. 7.7. Arora et al. (2017) inferred three major trends of the lineament orientations being NW–SE, N–S, and NNE–SSW. They correlated these trends with the subsurface fault pattern observed in the four boreholes around the survey area as well as subsurface magnetic anomalies (Arora et al. 2017). In the years to follow, the LiDAR data have been extensively used in comprehending the processes and certain important findings which have contributed immensely to our understanding of genesis of RTS at Koyna.

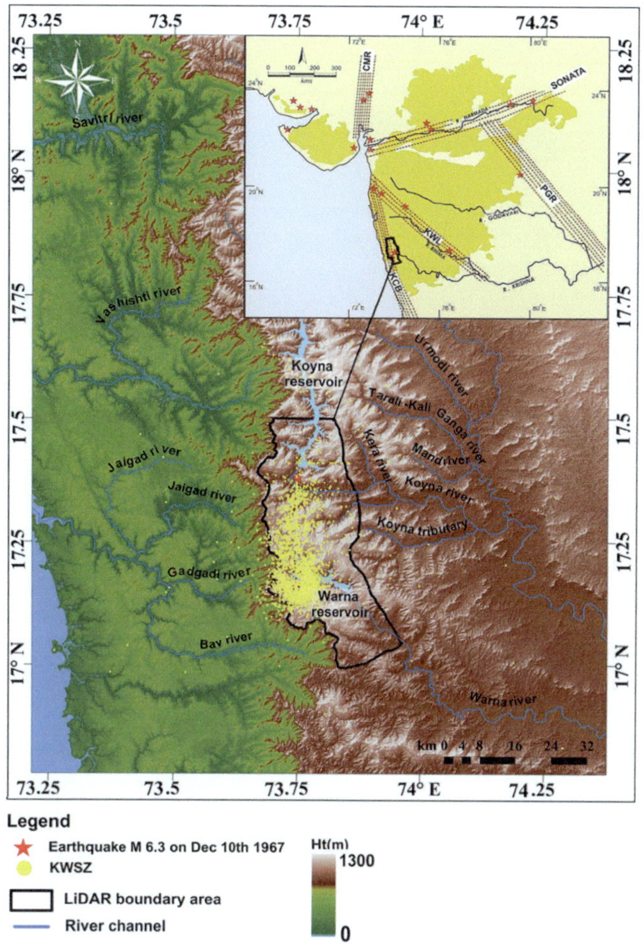

Fig. 7.6 The Koyna-Warna Seismic Zone (KWSZ) along Western Ghats Escarpment (WGE). Black outline shows airborne LiDAR survey coverage. Inset shows major tectonic feature along west coast and central India: CMR = Cambay Rift Zone, KCB = Konkan Coastal Belt, KWL = Kurduwadi Lineament, PGR = Pranhita Godavari Rift Zone, SONATA = Son Narmada Tapi belt, red stars = epicenters of M 4.0 earthquakes

7.6 Magnetotelluric Surveys

One of the most effective methods to delineate seismogenic zone in the Koyna region based on resistivity contrast due to faulting is application of the broad band magneto-telluric (MT) surveys. During 2012–2014, detailed MT surveys were conducted in the Koyna region. A total of 100 MT stations were established (Fig. 7.8). At the same time two remote sites were operated to improve the signal to noise ratio. MT data were acquired for 5 days at each of these 100 sites (Gupta et al. 2016). To improve signal to noise ratio, a few sampling frequencies were repeatedly recorded at different time intervals. A quantitative estimate of the nature of the variation of conductivity with depth was obtained from the

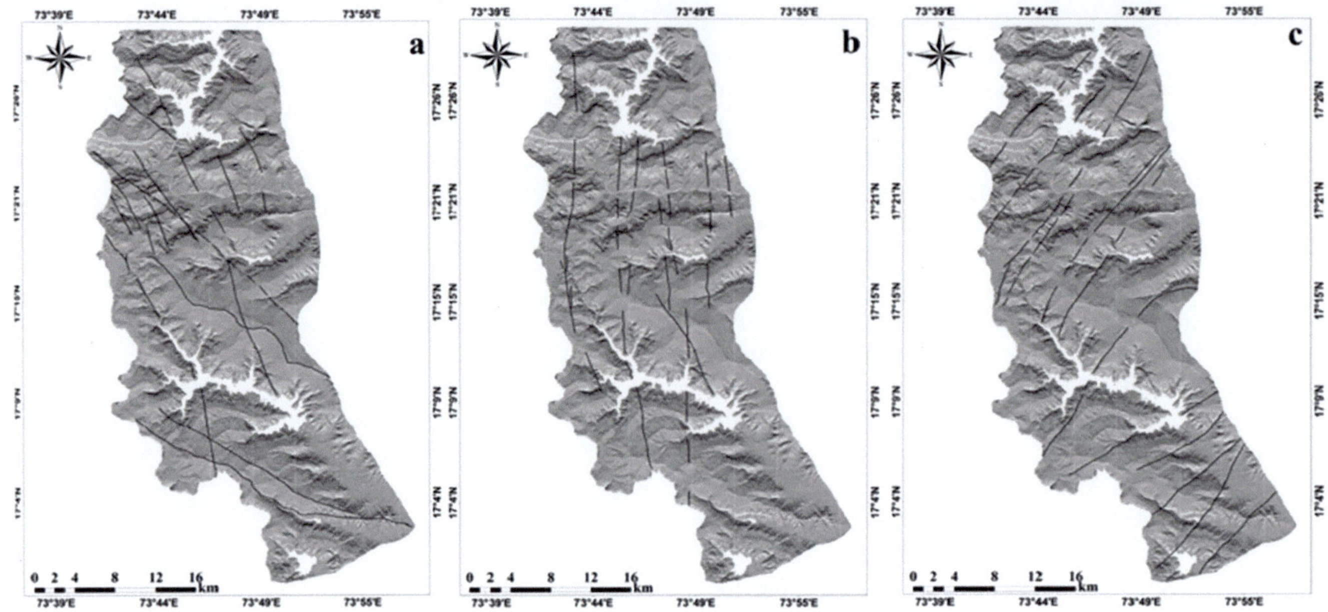

Fig. 7.7 Lineaments delineated from LiDAR high resolution DEM in three directions **a** NW–SE, **b** N–S and **c** NE–SW

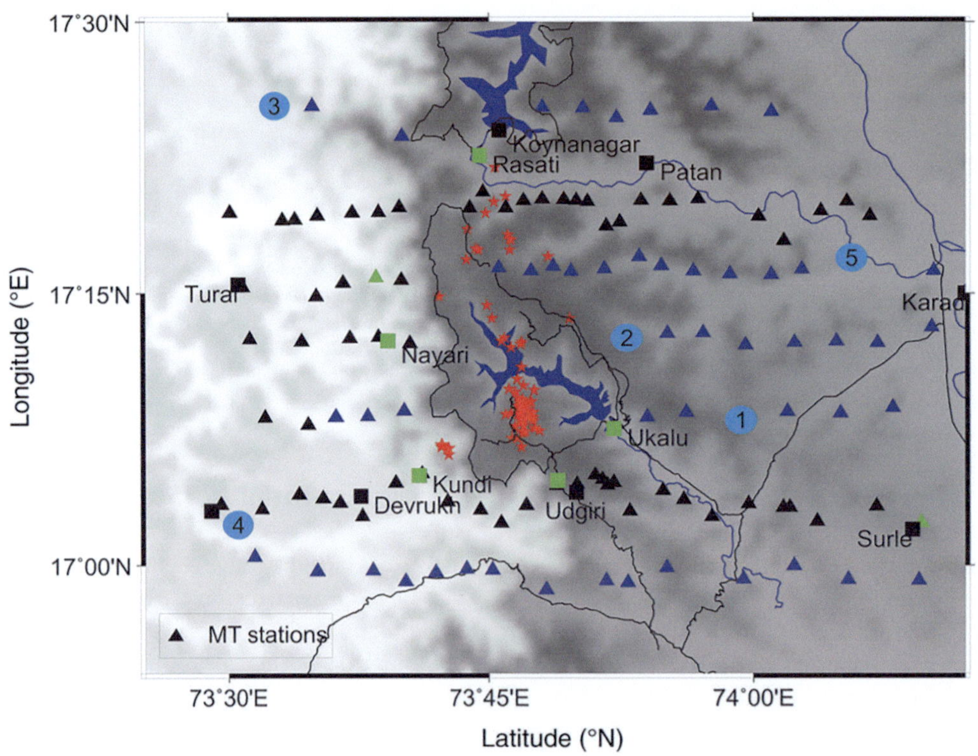

Fig. 7.8 Location of MT stations (triangles) recorded in the Koyna–Warna region. The numbered circles indicate the locations for which the transfer functions were performed. The squares indicate the name of the prominent towns in the area. The cluster of stars indicate earthquakes of M $\geq$ 3.0 from 2005 to 2015

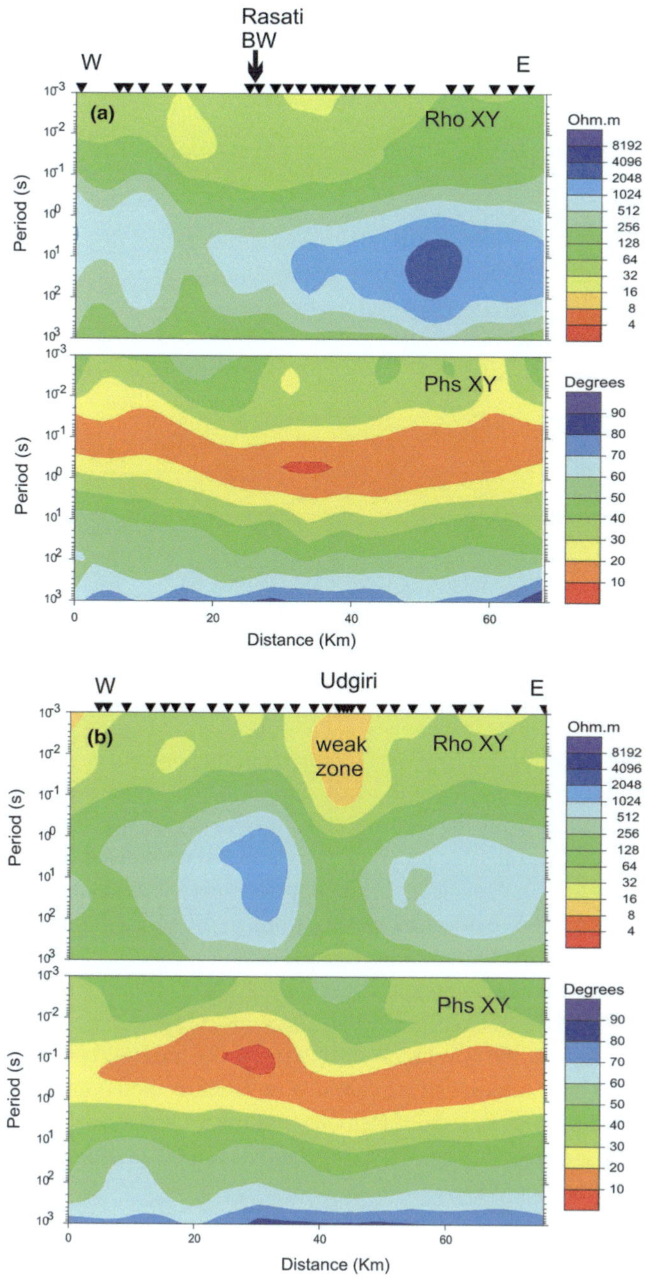

Fig. 7.9 Resistivity-phase pseudo-EW depth sections along MT-1 profile passing through Koyna (top) and MT-2 profile passing through Udgiri (bottom). The locations of these profiles are shown in Fig. 7.4

apparent resistivity and phase curves. Preliminary 1D and 2D modeling studies provided a broad picture of the Deccan Trap thickness, and the underlying basement configuration as depicted in Fig. 7.9. It is very interesting to note that the low resistivity anomaly extending to approximately 1 km depth was found to be true near the Udgiri, where a 1.5 km deep borehole was drilled and loose and less dense strata was encountered.

7.7 Heat-Flow Studies

Before the drilling of ~ 1.5 km deep boreholes in the Deccan Volcanic Province (DVP) under the integrated studies of RTS at Koyna, the heat flow measurements in the DVP were limited to a depth of ~ 200 m in the boreholes of opportunity. These measurements revealed a low heat flow of 41 m W m^{-2} as reported by Gupta and Gaur (1984). The subsurface temperature estimated from these studies was very poorly constrained. A borehole drilled to the depth of 1522 at Rasati (Fig. 7.10), in the immediate vicinity of the Koyna Dam was logged for temperature measurements. The borehole had penetrated the entire basalt pile of 933 m and went 589 m deep into the underlying granitic basement (Gupta et al. 2015). Thermal conductivity measurements were carried out on the cores obtained from the Rasati borehole yielded a heat flow of 45 m W m^{-2}. Temperature was estimated to be 130–150 °C at a depth of 6 km. Figure 7.10 shows the temperature depth profile in the Koyna area. This has been an important crucial estimate of the temperature while planning a deep bore hole (~ 7 km) research laboratory planned to be set up in Koyna (Roy et al. 2013a).

7.8 Geological Investigations

Detailed surveys were carried out in the rectangular block shown in Fig. 7.11. This block is in southern part of DVP. The entire RTS observed in the region is limited in the

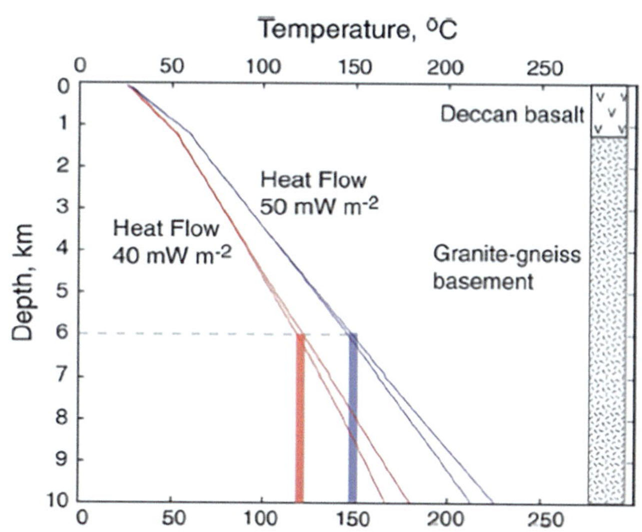

Fig. 7.10 Temperature-depth profiles in the upper crust in the Koyna area computed using a range of heat flow and heat production scenarios. Heat flow range is 40–50 m W m^{-2}. A low heat production (0.2 μ W m^{-2}) for basalt and a range of heat production for granite–gneiss (0.5–1.5 μ W m^{-2}) has been considered. Estimated temperature at 6 km depth is in the range of 130–150 °C

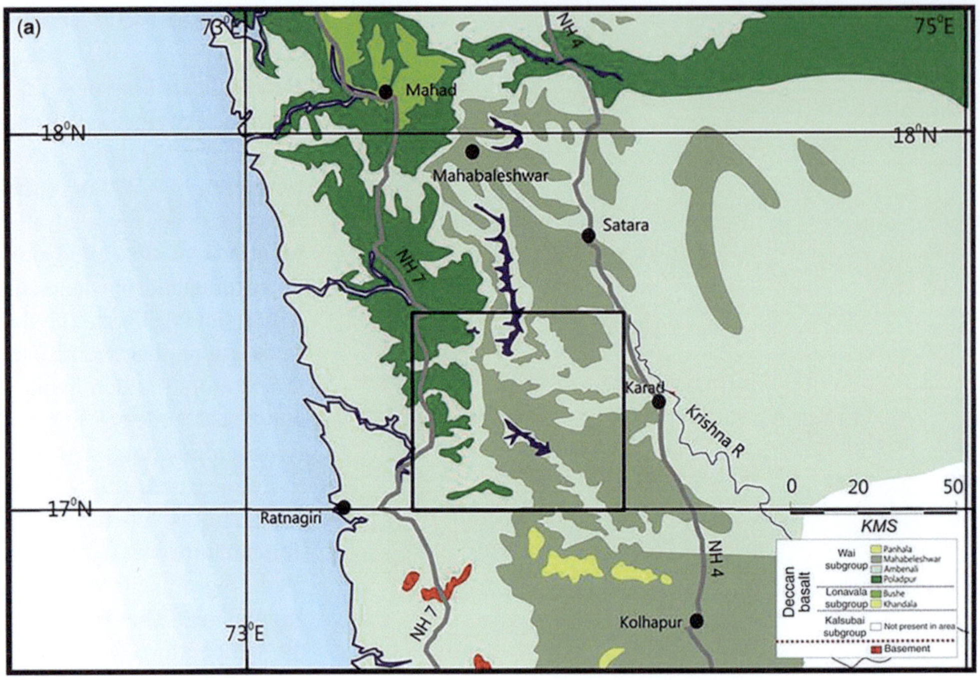

Fig. 7.11 Regional chemostratigraphy

north by Koyna reservoir and in the south by Warna reservoir (Fig. 7.11). The chemo-biostratigraphy of the region is depicted in Fig. 7.12. According to Walker (1971) the basaltic lava flows are simple (a'a-) type flows. The flows consist of amygdaloidal and vesicular basalts overlying compact basalt. At places the flows are topped by breccias or oxidized, ferruginous powdery horizons called 'bole' beds (Bondre et al. 2004). As can be seen from Fig. 7.12, the regional topography is dominated by the Western Ghats Escarpment (WGE), which has ~ 600 m level difference going from west to east. The boreholes drilled in the region are marked with white circles. Five of the boreholes, east of ESC, numbered 1, 2, 6, 7 and 8 are located east of WGE, on high elevation, while 3, 4, 5, and 9 are at lower elevations west of the WGE (Fig. 7.12). The locations of these boreholes were chosen to encompass the entire RTS area and provide a good azimuthal coverage. Boreholes # 1 and # 6 are located just south of the Koyna and Warna reservoirs. The availability of suitable location for drilling operations, without incursion into the forest areas, provided certain constraints (Gupta et al. 2016).

7.9 Well Logging in Koyna RTS Zone

The locations, elevation, depth and the thickness of the Deccan Trap for the 9 boreholes are provided in Table 7.4, and the logging details are provided in Table 7.5 (after Gupta et al. 2016). Considerable variations in the physical properties of the rocks in the boreholes was observed which are governed by the location of the borehole with respect to Western Ghats Escarpment (WGE). Basalt sections in the boreholes located at higher elevation, on the east side of WGE, show a marked difference from the boreholes located west of the WGE at lower elevations in caliper, gamma, and resistivity logs (Gupta et al. 2016). However, basement sections of the boreholes west of WGE exhibit more variations in the physical properties compared to boreholes located east of the WGE. This has been attributed to the heterogeneity of crustal material related to structural discontinuities/weak zones (Gupta et al. 2016). On average, the boreholes west of the WGE have 600 m lower elevation compared to the east side of WGE located boreholes, it is inferred that the basement column of these boreholes represent deeper part of the crust and bear the signature of the formation of WGE.

7.10 Donachiwada Source Fault

A preliminary delineation of the Donachiwada Fault was made through Soil-helium surveys in the surface rupture zones of the 1967 Koyna (M 6.3) stable continental region (SCR) earthquake (Gupta et al. 1999), as shown in Fig. 6.25 of Chap. 6. After the occurrence of the disastrous 1995 Kobe earthquake in Japan, the causative Nojima Fault was one of the first globally to have been investigated by putting 3 boreholes of 500, 800 and 1760 m depths and carrying out a suite of geological and geophysical investigations (Ando 2001). This helped in reconstructing the structure and evolution of the Nojima Fault. Other important examples of such investigations include the San Andreas Fault in California, USA (Hickman et al. 1994; Zoback et al. 2010) and the Chelungpu

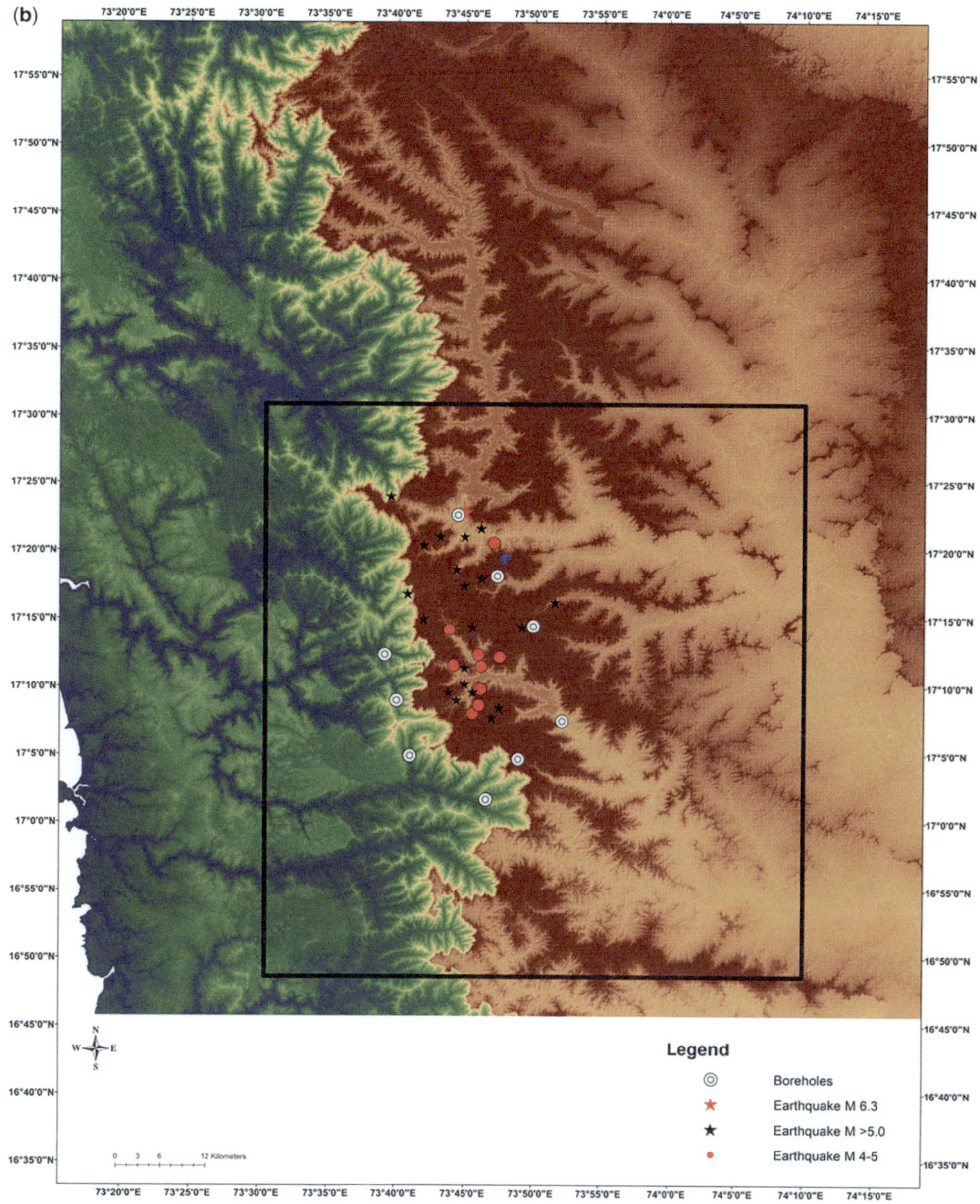

Fig. 7.12 Topography of the southwestern part of the DVP around the Koyna–Warna region; the box on each map refers to Fig. 7.11

Fault in Taiwan (Ma et al. 2006). For an in-depth study of the RTS a network of 20 surface seismic stations and 6 borehole stations had been set up. This has improved the location of epicenters to ± 300 m. Figure 7.13 updated from Gupta et al. (2016), shows the distribution of the broadband seismometers, borehole seismometers, and seismicity in the Koyna RTS region for the period August 2005 through December 2015. Figure 7.13 also depicts the surface trace of the Donachiwada Fault.

7.11 Major Scientific Findings in the 3 Years Following the 1st ICDP Workshop in 2011

Major findings of these 3-year-long investigations include (Gupta et al. 2014, 2016):

1. Discovery of near absence of the Mesozoic sediments below the basalt cover that had been long debated.

Table 7.4 Location, elevation, depth and Deccan Trap thickness in each borehole

Borehole name, locality	Latitude (° N)	Longitude (° E)	Elevation (m)	Final depth (m)	Thickness of Deccan Traps (m)	Remarks
KBH-1, Rasati	17.377	73.741	601	1522	932	Water loss
KBH-2, Udgiri	17.078	73.815	953	1196	1184	Significant thickness of soft formation
KBH-3, Kundi/Belari	17.082	73.683	124	1134	479	Presence of soft formation
KBH-4A, Nayari	17.206	73.653	98	1211	518	Disturbed basement
KBH-5, Phanasavle	17.150	73.667	161	906	499	Soft formations after transition
KBH-6, Ukhalu	17.125	73.869	581	1503	783	Highly fractured
KBH-7, Panchgani	17.302	73.789	949	1500	1251	Severe water loss
KBH-8, Paneri	17.241	73.833	977	1500	1198	Severe water loss
KBH-9, Khadi-Kolvan	17.028	73.776	145	1135	412	Warm water outflow

Table 7.5 Depths of conventional log data used in study

	Gamma ray logging depth	Guard resistivity logging depth	SP logging depth	Density logging depth	Caliper logging depth
KBH-1, Rasati	0–948		0–945		480–939
KBH-2, Udgiri	1–835		100–830		104–830
KBH-3, Kundi	90–1120		5–525		5–510
KBH-4A, Nayari	0–1178	784–1163	89–1178		89–1163
KBH-5, Phanasavle	114–555		105–920		
KBH-6, Ukhalu	115–1495	110–950	115–1495	789–1496	105–950
KBH-7, Panchgani	120–1480	120–1480	125–1495	120–1480	115–1490
KBH-8, Paneri	90–1475	250–1475	105–1485	90–1475	90–1485
KBH-9, Khadi-Kolvan	500–1133	110–1133	500–1133	100–1133	100–1133

2. Basalt cover thickness in the region is related to topography. While the surface topography has variations of ~ 1000 m, the basement is found to be flat at present lying – ~ 300 m below the current sea level.
3. The temperatures at a depth of 6 km were estimated to be 130 °C, indicating a rather cold and brittle crust. Earthquakes hypocenters concentrate around subsurface sharp density and resistivity contrast.
4. MT surveys delineated mechanically weak subsurface zones as well as tracing the basalt–basement contact.
5. Alignment of basement faults was revealed through the detailed geological/geophysical surveys.
6. LiDAR surveys helped in generating bare-earth model of the region as well as delineating the Donachiwada fault that hosted the 1967 M 6.3 and several M ~ 5 earthquakes.
7. Absolute location of the hypocenters was improved to ~ 300 m with the help of the network of the six borehole seismometers, from the earlier situation of having – 800 m uncertainties in the hypocenter locations.
8. Geological–geophysical logging in the nine boreholes was helpful in delineating the alignment of the faults.

It was important to discuss the work carried out in during 2011–2014 with the international community. A second ICDP Workshop was held at Koyna during 16–18 May 2014. It was attended by 49 participants, several of them having attended the 1st ICDP Workshop of 2011. The participants had detailed field visits and discussed at length the work having been carried out during 2011–2014 (Gupta et al. 2014). The scientific questions discussed, several of them having been taken from the San Andreas Fault Observatory at Depth (SAFOD) program (Zoback et al. 2011) included:

1. What is the fluid pressure and permeability within and adjacent to the fault zone?
2. What are the composition and origin of fault-zone fluids and gases?
3. How do stress orientation and magnitude vary across fault zones?
4. How do earthquakes nucleate?
5. How do earthquake ruptures propagate?
6. How do earthquake source parameters scale with magnitude and depth?

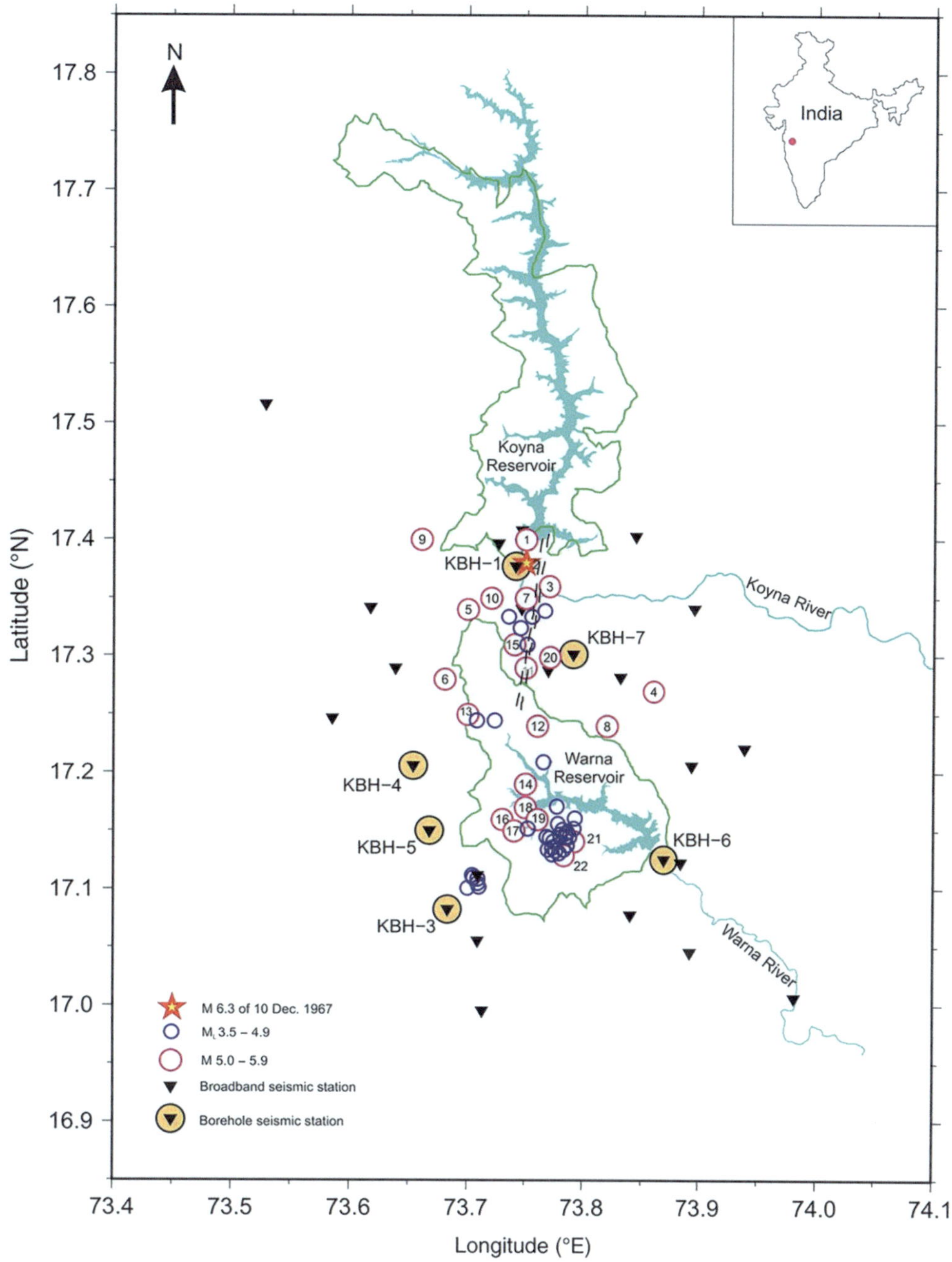

Fig. 7.13 Map of the Koyna–Warna region, western India, showing the distribution of broadband seismic stations, epicenters (M 3.5–4.9 during August 2005–December 2015; M ≥ 5.0 since 1967 numbered in chronological order), and locations of the six borehole seismic stations. The green curve is the reserved forest boundary. Dashed lines (black) indicate the Donachiwada fault zone (Gupta et al. 1999). The inset indicates the study area in India (updated from Shashidhar et al. 2016)

7. What is the role of water reservoirs in triggering earthquakes?
8. What is the 3-D/4-D nature of the fault zone?

It was observed that the geological, hydrological and geophysical work carried out in the Koyna region provides a good framework to study the regional tectonics; however, it lacks the critical inputs required to explore the physical mechanism that connects the earthquake occurrences with the reservoir-level changes. The proposed scientific deep drilling and setting up of a deep borehole laboratory to study pre-seismic, co-seismic and post-seismic changes in the physical properties of the fault and the rocks in its vicinity are desirable. It was realized that by instrumenting the deep borehole for long-term monitoring of the critical parameters such as seismicity, temperature, fluid/gas and pore pressure would provide unprecedented critical near-field information on the temporal changes in these parameters before, during and post occurrence of earthquakes. The participants of the 2nd ICDP Workshop were happy with the progress made over the last three years (2011–2014) and gave a go ahead with drilling of the Pilot Borehole(s) (Gupta et al. 2014). A drilling plan was also presented and discussed in detail.

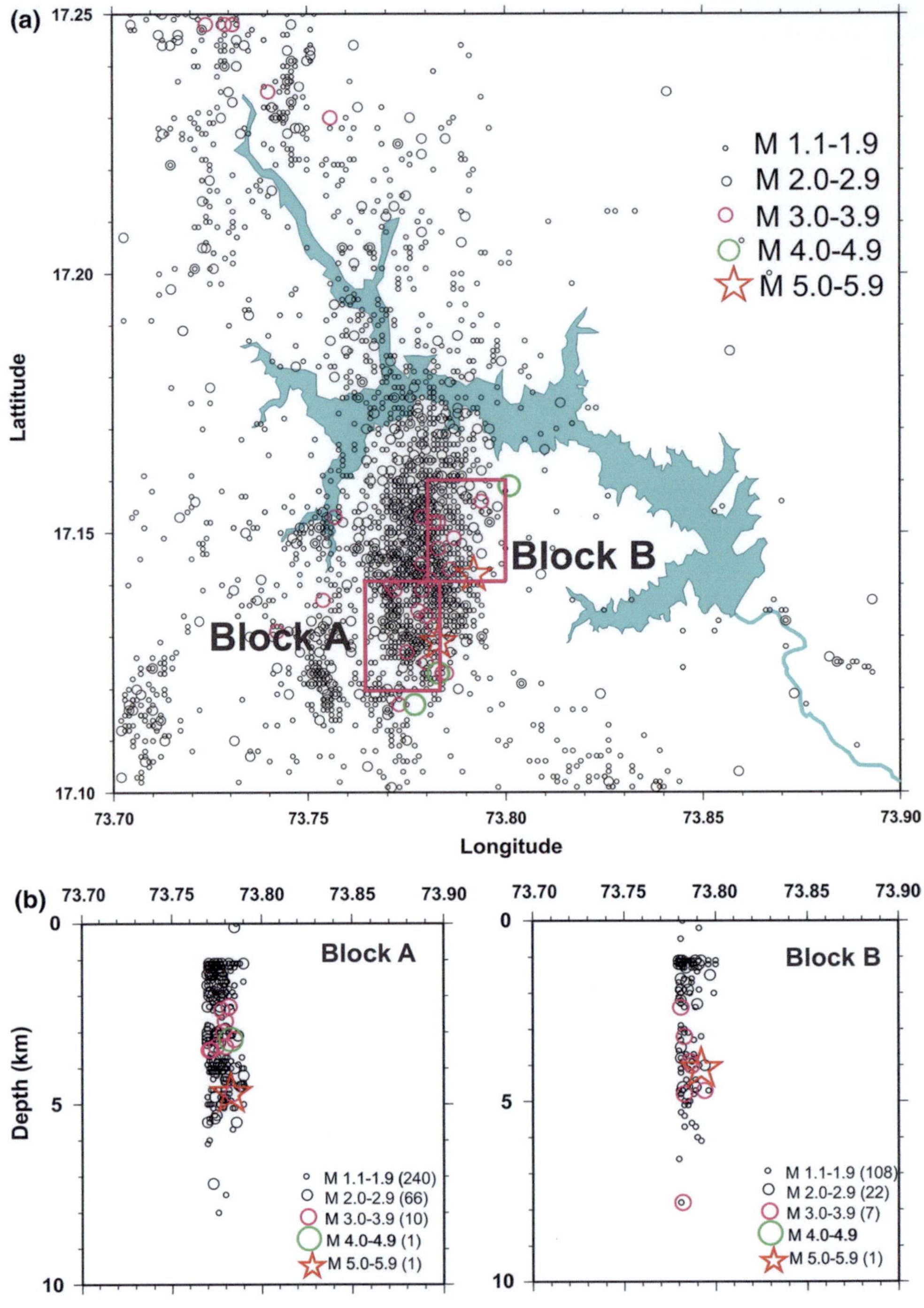

Fig. 7.14 **a** Seismicity in the vicinity of the Warna reservoir during the study period from 2009 to April 2014. The two blocks A and B are each 2 km × 2 km in area. **b** Focal depth sections in the two blocks A and B within the Warna seismic cluster shown in **a**

7.12 Location of the Pilot Borehole

After the very successful second ICDP Workshop held during 2014, the next step was to find the most suitable location for a 3 km deep Pilot Borehole. The chosen Pilot Borehole location had to be easily accessible, be suitable for drilling, and required to be in the vicinity of the Donachiwada Fault which had hosted the 1967 M 6.3 and several ~ M 5 earthquakes. The available seismicity maps for the Koyna RTS area were examined in detail. Figure 7.14 depicts seismicity in the vicinity of the Warna Reservoir for the period 2009 through April 2014. Two blocks, A and B, 2 km × 2 km in area were considered (Fig. 7.14). (a) The depth distribution of earthquakes of M > 1, during 2009 through 2014 is depicted in Fig. 7.14. (b) There are plenty of earthquakes within this small area, including an M 5 magnitude earthquake, mostly confined to a depth of 1–6 km. Three borehole locations, south to the Koyna Reservoir, were also considered.

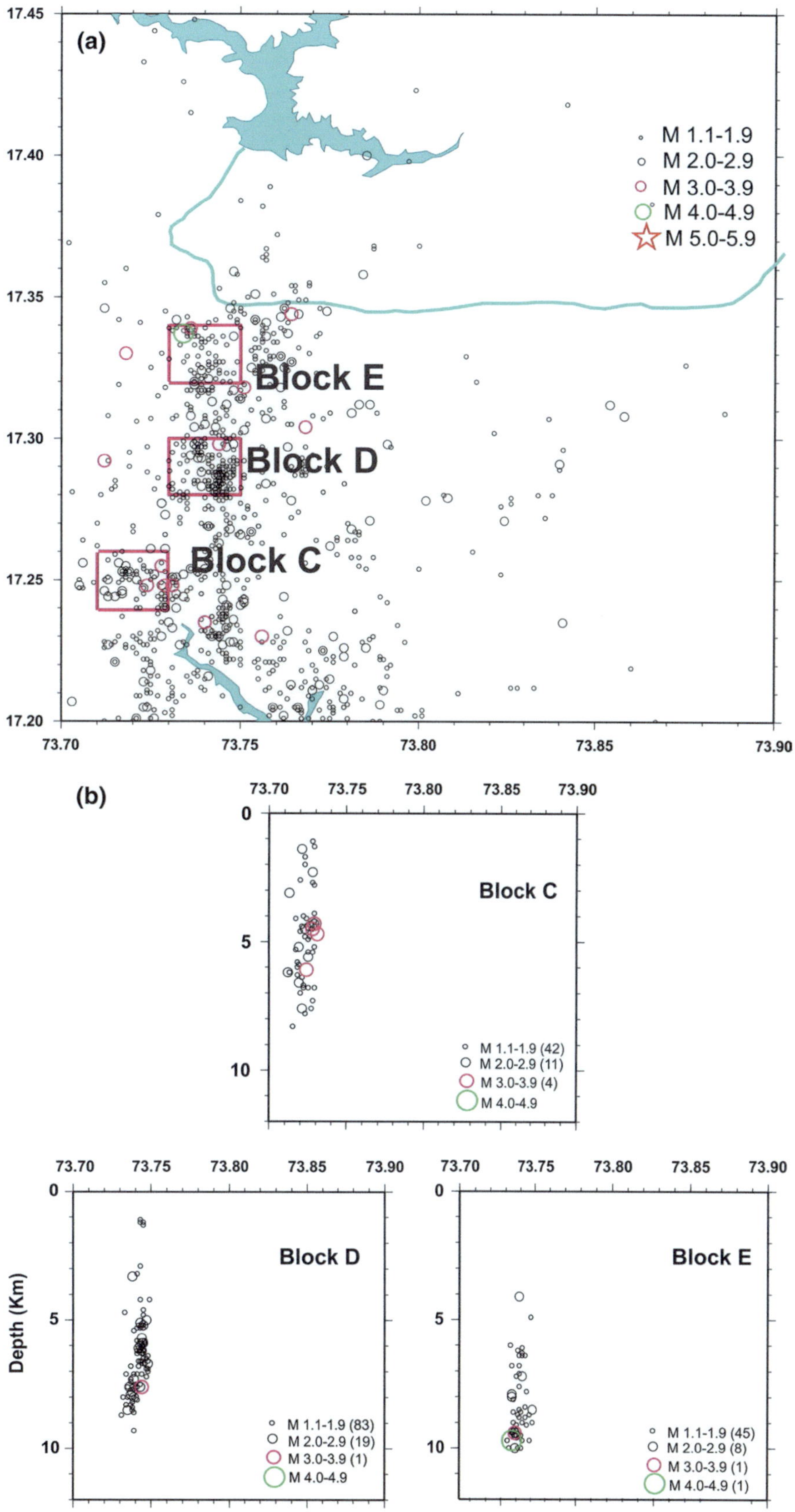

Fig. 7.15 **a** Seismicity in the Koyna region during 2009–April 2014. The three blocks C, D and E are each 2 km × 2 km in area. **b** Focal depth sections in the three blocks C, D and E shown in **a**

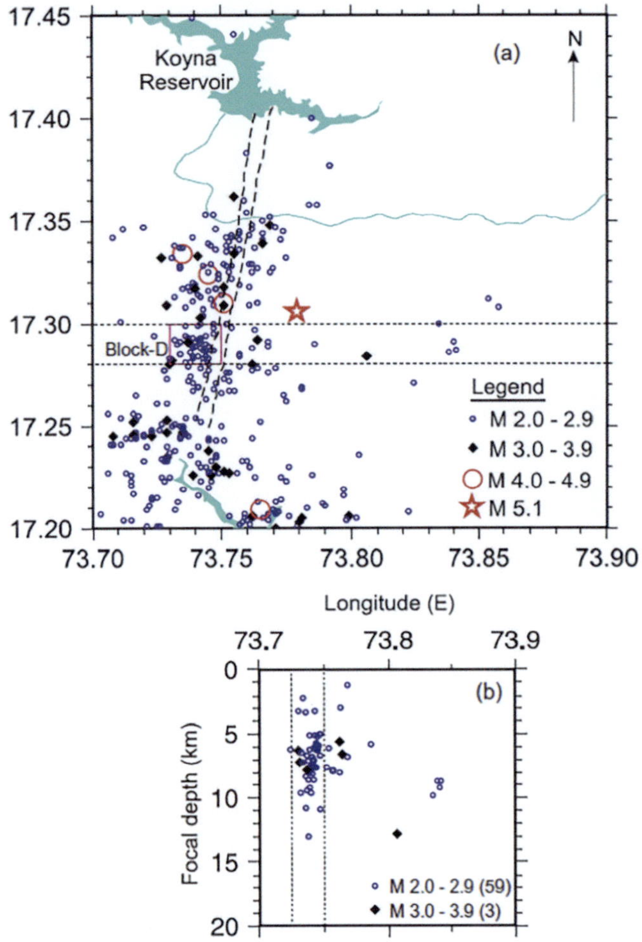

Fig. 7.16 **a** Earthquakes of magnitude $\geq$ 2.0 during August 2005–December 2015. Dotted lines indicate a swath of 2 km area of the block D. Dashed line indicates the Donachiwada fault zone (Gupta et al. 1999). **b** Depth section for the 2 km wide swath in **a**. Dotted lines indicate the area of 2 × 2 km^2

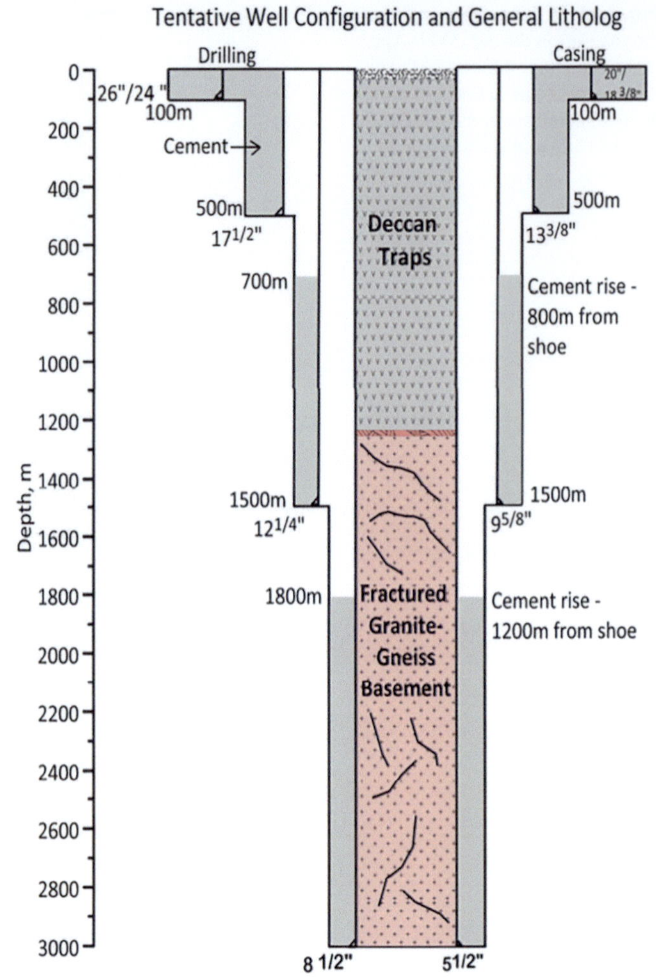

Fig. 7.17 Generalized configuration of the 3000 m deep pilot borehole, giving the drilling and casing details. The borehole entered the basement at a depth of 1247 m after penetrating through the basalt column. Practically no sediments were encountered below the basalt column, which is lying directly on the pre-Cambrian granite-gneiss basement

Figure 7.15a shows the location of these three blocks, C, D, and E and the epicenters of the earthquakes that occurred during January 2009 through April 2014. Figure 7.15b exhibits the depth distribution of the hypocenters in the three chosen blocks of 2 km × 2 km. Other than block C, where the focal depths range from 1 to 8 km, in blocks C and D, there is a concentration of the hypocenters in depth range of 3–10 km. Figure 7.16a shows the location of the block D, Donachiwada Fault as deciphered earlier (Gupta et al. 1999), and earthquake epicenters of M $\geq$ 2 earthquakes for the period August 2005 through December 2015. Figure 7.16b shows the depth section of the earthquakes that occurred in a 2 km wide swath marked in Fig. 7.16a. Keeping logistics and vicinity to the Donachiwada Fault, block D was finally chosen to place the 3 km deep Pilot Borehole.

7.13 Drilling and Completion of the 3 km Deep Pilot Borehole

The spudding for the 3 km deep Pilot Borehole was done on 20 December 2016. A depth of three km was reached on the 11 June 2017. General lithology and the well configuration of the Pilot Borehole is provided in Fig. 7.17. The details of the progress of drilling are depicted in Fig. 7.18. The basement was reached at a depth of 1247 m. It is to be noted that the nearby Panchgani borehole, the basement was met at a depth of 1252 m. This provided support to the inference that the basement in the region is more or less flat at an average depth of 300 m below the present sea level. There were no sediments below the basalt cover, which was underlain by the

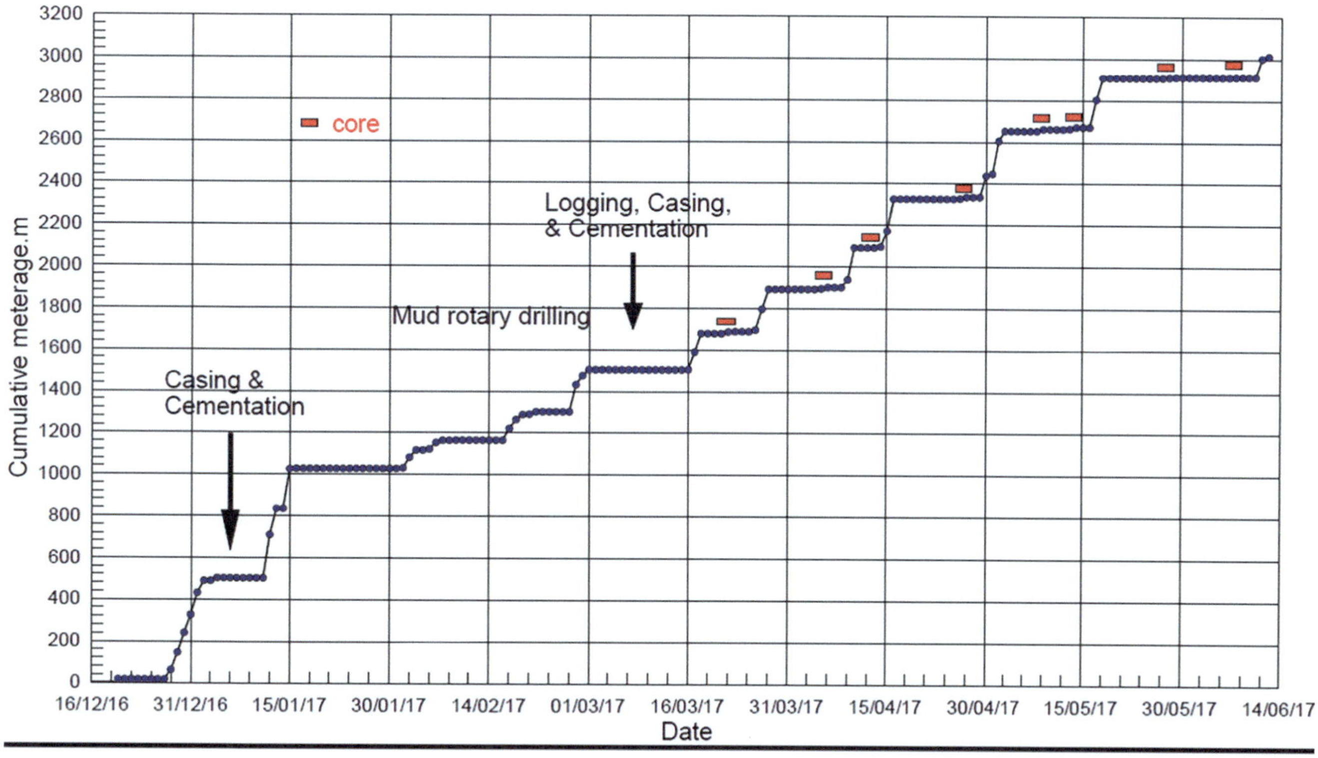

Fig. 7.18 Progress of drilling KFD-1

granitic basement. While drilling, several zones were encountered where there was immense loss of the drilling mud, indicating the presence of the faults being intersected by the borehole. There was detailed geophysical logging for the entire borehole. Drilling was a combination of hammer and rotary drilling. On-Line-Gas-Analysis (OLGA) facilities had been set up in collaboration of ICDP that provided valuable information. Cores were collected at the depths of 1679, 1892 and 2091 m depths through rotary drilling. These were of 4 in. diameter and up to 9 m in length. In-situ-stress measurements were conducted at 1600 m and deeper depths.

A Post-Operational ICDP workshop was held at Pune, India during 14–16 October 2017. It was attended by 67 participants including 19 from Canada, France, Germany, Italy, Japan, the Czech Republic, South Korea, Spain, Switzerland, and the USA. There was a total satisfaction with the results obtained and plans of further work leading to designing the proposed 7 km deep Main Borehole. Participants of the workshop visited Borehole Geophysics Research Laboratory at Karad. The Core Repository was of immense interest. On the final day, detailed recommendations for the future course of action were made focusing on developing of Koyna RTS information system; characterization of Donachiwada fault; deployment of a vertical string of seismometers below 1500 m in the Pilot Borehole; a variety of hydrogeological/geophysical/geological measurements to be made; detailed study of the core samples; setting up of a reference borehole outside Koyna RTS area; and preparation of a detailed plan of the proposed deep borehole laboratory at a depth of ~ 7 km. Overall, the placing of the borehole at a very appropriate location and carrying out all the necessary measurements during the drilling of 3-km-deep borehole were appreciated.

The work was going on in full swing after the third ICDP workshop when the COVID-19 pandemic put brakes on the progress and broke the momentum. During the last 3 years of uncertainty that is continuing has resulted in certain setbacks in meeting the planned targets. Some of the planned targets and the timelines were worked out keeping in view the conditions prevailing in the country and in the world. I am hopefully, that a dream of drilling a 7 km deep borehole right into the heart of the seismic zone to catch a *Live* earthquake may become a reality someday.

7.14 Continued RTS in the Koyna Region

As mentioned earlier, no earthquakes were reported from the Koyna region prior to the impounding of the Shivaji Sagar Lake, created by the Koyna Dam in 1962. Gupta et al. (2002) debated as to how long the RTS will continue in the Koyna region. It is hypothesized that Koyna region was stressed close to critical at the time of the impoundment of the Shivaji Sagar Lake by the construction of the Koyna Dam and it could host a M 6.8 earthquake (Johnston et al. 1994).

By the year 2002 the total number of $M \geq 5$ earthquakes were 17 and there were some 170 M ~ 4 earthquakes and of course several hundred smaller earthquakes. Gupta et al. (2002) estimated that about one-half of an M 6.8 magnitude energy has been already released. They also estimated that the remaining stored strain energy would be released in the next 3 decades. They also hypothesized that there would be no repeat of the 10 December 1967 earthquake. And the occurrence of M 5 earthquakes would be governed by Kaiser (1953) effect. As of the end of 2024 there have been 22 $M \geq 5$ and some 200 $M \geq 4$ earthquakes. Following is the estimate of percentage of a 6.8 magnitude energy released by the end of 2024:

The Max. Credible Earthquake (M_{CE}) considered for Koyna: M = 6.8

Empirical relation used: Log E = 1.5 M + 11.8 (M = magnitude)

Using the above relation $E_{MCE} = 10^{22}$

Considering the 10 December 1967 M 6.3, 22 M ~ 5.5 and 200 M 4, the energy released so far is estimated to be: $E = 1 \times E_{M\,6.3} + 22 \times E_{M\,5.5} + 200 \times E_{M\,4.5}$

$$E = 10^{(21.25)} + 22 \times 10^{(20.05)} + 200 \times 10^{(18.55)} = 10^{21.94}$$

This amounts to some 87% energy release of an M 6.8 earthquake (percentage of $M_{CE} = 10^{21.94}/10^{22} = 87\%$). It may hence be stated that most of the energy of a M 6.8 earthquake has been released over the years in the Koyna region. Table 7.6 updated from Arora et al. (2017) provides a list of $M \geq 4$ earthquakes, in the Koyna region from 1967 through 2024. From this list, it may be noted that no $M \geq 4$ earthquakes has occurred in the Koyna region after the 14 December 2021 earthquake of M 4.0. The gradually decreasing RTS at Koyna is also depicted in Fig. 7.19 updated from Anuradha and Mallika (2017). This figure shows the number of $M \geq 4$ earthquakes per annual loading/unloading cycle of Shivaji Sagar Lake created by Koyna Dam. Usually, following the Monsoon season during the months of June–September, the Shivaji Sagar Lake gets filled up with the inflow of rainwater and the peak levels are reached during August–September. The highest level of RTS is observed in the month of September, as depicted in Fig. 7.20, where the monthly number of $M \geq 4$ are plotted for the period 1967 through 2024 (updated from Gupta 2017). It may also be noted from this figure that after September, the second most active month is December. Emptying of the reservoir follows the rainy season and the lowest levels are reached in the months of May and June. The emptying also gives rise to RTS with another peak RTS being observed in the month of March (Fig. 7.20).

Arora et al. (2017) have shown that most of the $M \geq 4$ RTS events in the Koyna region have occurred in a 10 km wide and 35 km long zone extending north to south running between the Koyna and Warna Reservoirs (Fig. 7.21 updated from Arora et al. 2017). They have demarcated this high RTS zone on the LiDAR based Digital Elevation Model. Another interesting observation made by them is a plot of month wise distribution of $M \geq 4$ earthquake epicenters. In Fig. 7.22, updated from Arora et al. (2017), the earthquake of $M \geq 4$ are plotted month-wise for the period 1967 through 2021. It must be mentioned that no $M \geq 4$ earthquakes has occurred in the Koyna region after 15 November 2021. This figure depicts that a majority of $M \geq 4$ earthquakes in the Koyna region during 1968 through 1985 occurred following the peak reservoir levels reached in the Shivaji Sagar Lake. A decrease in the seismic activity in the Koyna region is seen during 1986 through 1992. Following the impoundment of the Warna reservoir in 1993, an increase in seismicity is observed during the months of March–April in addition to the month of September.

7.15 Changes in '*b*' Values and RTS at Koyna-Warna

It has been reported amply in literature that *b* values associated with RTS sequences are considerably lower than the regional *b* values (Gupta 1992). In an interesting study, Mallika et al. (2013) have investigated the temporal variation of *b* values associated with M ~ 4 earthquake in the Koyna-Warna region. They specifically analyzed the data of earthquakes occurring in the vicinity of the Warna reservoir. They have used the catalog of ~ 3000 earthquakes that occurred in the vicinity of Warna region during August 2005 through December 2010. Table 7.7 gives the number of earthquakes in different magnitude ranges during August 2005 through December 2010 and the associated *b* values. The temporal variation of b values, specifically associated with $M \geq 4$ earthquakes in the Warna region was investigated. Mallika et al. (2013) used the moving window approach to investigate the variation in *b* values using a window length of 30 days and a time shift of 5 days. The computed *b* values are plotted as a function of time with the occurrence of $M \geq 4$ earthquakes as depicted in Fig. 7.23. It is noteworthy that a drop in *b* values is observed preceding all the 14 among 16 earthquakes of $M \geq 4$ (Table 7.8). The two earthquakes that are associated with high *b* values of 1.30 and 1.34 occurred on 17 April and 21 May 2006. Perhaps the reason of this difference is the occurrence of these two earthquakes associated with the emptying of the Warna reservoir, while the remaining 14 earthquakes occurred during the loading phase of the Warna reservoir. Mallika et al. (2013) conclude that closely observing the temporal variations in the *b* values is a good indicator of the possible occurrence of $M \geq 4$ earthquakes in the vicinity of Warna reservoir.

Table 7.6 Catalogue of M ≥ 4.0 earthquakes from 1967 through 2024 compiled from various sources (Gupta et al. 1980; Gupta 2002; Talwani et al. 1996)

S. No.	Year	Month	Date	Origin time	Mag.	Depth (km)	Lat.	Long.
1	1967	9	13	6:23	5.8			
2	1967	9	13	6:47	4.5			
3	1967	9	13	6:51	4.0			
4	1967	9	13	7:01	4.0	5	17.348	73.787
5	1967	9	13	8:43	4.0			
6	1967	9	13	11:32	4.0			
7	1967	12	10	22:51	6.3	10.36	17.347	73.748
8	1967	12	11	20:49	5.4			
9	1967	12	12	18:20	4.7	15	17.234	73.757
10	1967	12	13	5:09	4.6	30.36	17.266	73.878
11	1967	12	13	19:19	4.6	27.94	17.411	73.935
12	1967	12	14	9:16	4.1	15	17.311	73.777
13	1967	12	14	15:06	4.1	6.19	17.287	73.799
14	1967	12	14	23:40	4.0	4.65	17.274	73.770
15	1967	12	22	14:48	4.8			
16	1967	12	24	3:41	4.0	26.8	17.277	73.748
17	1967	12	24	4:23	4.0	4.99	17.208	73.581
18	1967	12	24	23:49	5.0	15	17.202	73.813
19	1967	12	25	0:15	4.2	19.94	17.232	73.804
20	1967	12	25	0:47	4.2	15	17.248	73.748
21	1967	12	25	17:37	4.6	8.4	17.330	73.734
22	1967	12	25	17:59	4.1	5.08	17.293	73.780
23	1968	1	3	4:35	4.0	3.53	17.351	73.737
24	1968	1	11	4:37	4.1	0.63	17.409	73.771
25	1968	1	16	3:33	4.0	15	17.303	73.816
26	1968	2	7	8:09	4.3	8.23	17.294	73.733
27	1968	2	9	22:52	4.2	9.06	17.250	73.804
28	1968	2	12	9:13	4.5			
29	1968	3	4	21:36	4.3	6.19	17.2902	73.8145
30	1968	8	31	2:53	4.5	2.23	17.3873	73.6893
31	1968	9	20	10:11	4.4	5.76	17.4100	73.7713
32	1968	10	29	10:00	5.0	1.05	17.3700	73.7480
33	1968	10	30	11:29	4.1			
34	1968	11	23	11:12	4.1			
35	1969	1	21	22:32	4.1			
36	1969	2	13	18:26	4.2	8.04	17.3802	73.6847
37	1969	3	7	14:28	4.4	4.17	17.2797	73.7260
38	1969	6	3	23:26	4.2			
39	1969	6	27	20:05	4.5	3.53	17.3720	73.7923
40	1969	7	22	21:49	4.0			
41	1969	11	3	23:22	4.4	9	17.3202	73.7480
42	1969	11	4	5:11	4.2			

(continued)

Table 7.6 (continued)

S. No.	Year	Month	Date	Origin time	Mag.	Depth (km)	Lat.	Long.
43	1970	4	16	14:46	4.0			
44	1970	5	27	12:45	4.8	3.52	17.3710	73.7480
45	1970	6	8	5:30	4.1			
46	1970	6	17	6:48	4.1			
47	1970	9	21	3:02	4.0			
48	1970	9	25	4:12	4.6	6.94	17.4172	73.7958
49	1970	9	26	16:36	4.6	7.12	17.2823	73.7585
50	1971	1	23	4:39	4.2			
51	1971	2	14	1:30	4.0	3.68	17.3402	73.7760
52	1971	8	10	4:30	4.0			
53	1971	8	10	6:15	4.3	13.63	17.2950	73.7480
54	1971	11	22	10:39	4.7	15	17.3802	73.7248
55	1972	5	1	21:11	4.2			
56	1972	5	11	11:49	4.5			
57	1972	7	5	23:17	4.0			
58	1972	11	11	4:02	4.1			
59	1973	4	19	8:45	4.1			
60	1973	10	17	8:03	4.0			
61	1973	10	17	14:13	4.1	5	17.3900	73.7520
62	1973	10	17	15:24	5.0	4.05	17.3735	73.7812
63	1974	2	17	14:06	4.5	8.02	17.1632	73.7480
64	1974	4	28	9:30	4.0			
65	1974	5	29	18:26	4.2			
66	1974	7	29	23:17	4.8	19.01	17.3208	73.8395
67	1974	8	7	4:23	4.1			
68	1974	8	28	20:20	4.5	5.03	17.2072	73.7517
69	1974	11	11	15:11	4.3	5.96	17.2260	73.7878
70	1974	12	20	14:16	4.0	5	17.3628	73.7517
71	1975	2	10	18:35	4.0	5	17.3537	73.7965
72	1975	9	2	23:17	4.2	5.97	17.3390	73.7313
73	1975	12	2	7:40	4.2	0.13	17.2368	73.6870
74	1975	12	24	13:25	4.3	6.87	17.2213	73.6775
75	1976	3	14	5:16	4.8	3.98	17.2438	73.7517
76	1976	4	22	10:46	4.3	0.03	17.2802	73.6583
77	1976	9	16	14:04	4.2			
78	1976	9	26	6:48	4.3			
79	1976	12	12	0:52	4.2	5	17.3040	73.7517
80	1977	9	19	0:03	4.7	5.56	17.3230	73.8035
81	1977	11	4	18:56	4.0			
82	1977	11	4	20:36	4.2			
83	1977	11	4	20:55	4.4			
84	1978	11	29	5:30	4.0			
85	1978	12	12	15:01	4.4	5.47	17.2210	73.7845
86	1979	1	26	19:12	4.0			

(continued)

Table 7.6 (continued)

S. No.	Year	Month	Date	Origin time	Mag.	Depth (km)	Lat.	Long.
87	1979	9	26	20:02	4.0			
88	1980	2	6	22:13	4.6	6.82	17.1508	73.7238
89	1980	8	19	22:32	4.2			
90	1980	9	2	16:39	5.3	14.02	17.1740	73.8378
91	1980	9	2	16:47	4.5			
92	1980	9	20	7:28	5.5	15	17.1408	73.7390
93	1980	9	20	10:45	5.8	15	17.1543	73.6755
94	1980	9	20	11:22	4.5			
95	1980	9	20	14:27	4.0			
96	1980	9	20	23:44	4.2			
97	1980	9	21	0:00	4.0			
98	1980	9	21	3:52	4.0			
99	1980	9	21	18:19	4.0			
100	1980	9	22	11:59	4.3			
101	1980	9	25	13:38	4.3			
102	1980	9	27	8:54	4.0			
103	1980	9	30	13:37	4.2			
104	1980	10	3	15:20	4.4			
105	1980	10	4	16:37	5.1	5.52	17.2020	73.7575
106	1980	10	4	19:10	4.3			
107	1980	10	5	16:10	4.0			
108	1980	10	16	21:26	4.1			
109	1980	10	17	21:47	4.1			
110	1980	10	21	5:02	4.2			
111	1980	10	26	1:32	4.5			
112	1981	1	25	20:30	4.0			
113	1982	4	25	23:04	4.4	7.09	17.2017	73.7425
114	1982	5	5	7:32	4.2	5.01	17.2395	73.7212
115	1982	9	10	2:42	4.4			
116	1983	2	5	22:52	4.3	2.18	17.3047	73.6897
117	1983	3	21	15:02	4.1	6.53	17.2185	73.7698
118	1983	5	13	5:53	4.1	6.22	17.2612	73.7252
119	1983	5	28	18:08	4.2	21.77	17.2958	73.7403
120	1983	9	25	18:55	4.8	0.51	17.2240	73.8518
121	1983	11	10	8:55	4.5			
122	1984	9	25	7:47	4.6	15	17.5470	73.7852
123	1984	11	14	11:58	4.7	8.2	17.2375	73.7795
124	1984	12	21	17:26	4.1	8.69	17.2228	73.7755
125	1985	5	27	6:57	4.1	9.51	17.2397	73.7055
126	1985	10	29	8:33	4.0			
127	1985	10	29	13:59	4.2			
128	1985	11	15	7:02	4.3			
129	1985	11	21	9:28	4.0			

(continued)

Table 7.6 (continued)

S. No.	Year	Month	Date	Origin time	Mag.	Depth (km)	Lat.	Long.
130	1985	11	21	11:39	4.1			
131	1985	12	15	13:11	4.3	7.36	17.3627	73.7625
132	1985	12	28	14:52	4.0	20.45	17.3618	73.7480
133	1988	7	24	5:33	4.8	15	17.1992	73.7847
134	1988	8	15	22:16	4.0	7.3	17.2582	73.7297
135	1988	8	15	23:27	4.1	9.29	17.3652	73.6893
136	1988	9	11	20:39	4.4	12.68	17.2155	73.8135
137	1989	10	29	7:30	4.0	7.1	17.3700	73.7847
138	1990	7	3	13:28	4.1			
139	1991	1	6	22:13	4.4	6.27	17.1973	73.7355
140	1992	2	20	5:42	4.3	7.59	17.2583	73.6927
141	1992	4	1	1:34	4.0	5.77	17.3198	73.7322
142	1993	8	27	22:12	4.1	2.41	17.2147	73.7775
143	1993	8	28	4:27	5.3	11.92	17.2187	73.7798
144	1993	8	28	8:31	4.0	15	17.2107	73.7908
145	1993	9	3	23:03	5.0	13.75	17.2090	73.7537
146	1993	9	4	0:53	4.3	1.23	17.1958	73.7627
147	1993	10	22	1:15	4.1	11.01	17.1327	73.7518
148	1993	10	22	15:35	4.1			
149	1993	12	8	1:42	5.2	8.23	17.2050	73.7632
150	1993	12	21	10:10	4.1	12.15	17.1603	73.7587
151	1994	1	22	11:12	4.0	8.11	17.1882	73.7475
152	1994	2	1	9:31	5.5	8.83	17.3235	73.7480
153	1994	3	29	7:51	4.3	6.55	17.2080	73.7107
154	1994	10	31	6:10	4.2	9.93	17.2182	73.7735
155	1995	3	12	8:22	5.1	7.54	17.2457	73.7288
156	1995	3	13	3:09	4.9	5.37	17.2222	73.7228
157	1995	3	20	6:58	4.1	5.22	17.2390	73.7193
158	1995	4	15	22:59	4.2	8.49	17.2178	73.7167
159	1995	9	5	0:54	4.0			
160	1995	11	8	20:06	4.1			
161	1995	11	15	4:08	4.1			
162	1996	4	26	12:19	4.6			
163	1996	4	26	12:36	4.2			
164	1997	4	25	16:22	4.3	5	17.3430	73.7580
165	1998	2	11	1:08	4.7			
166	1998	2	14	0:59	4.4			
167	1999	6	7	15:45	4.3	3	17.3020	73.7690
168	2000	3	12	18:03	5.6	15	17.2000	73.7200
169	2000	3	13	20:46	4.0			
170	2000	4	6	22:30	5.1	8	17.1370	73.7500
171	2000	4	27	23:21	4.1	5	17.1630	73.7330
172	2000	9	5	0:32	5.6	10	17.2150	73.7750

(continued)

Table 7.6 (continued)

S. No.	Year	Month	Date	Origin time	Mag.	Depth (km)	Lat.	Long.
173	2000	12	8	13:23	4.1	5	17.1480	73.7490
174	2001	3	19	16:38	4.0	8	17.1900	73.7340
175	2001	5	17	16:04	4.4	8	17.1890	73.7350
176	2001	8	2	4:09	4.2			
177	2003	3	22	1:00	4.0			
178	2003	3	27	6:18	4.2			
179	2005	3	14	9:43	5.2	13	17.1000	73.7900
180	2005	3	15	2:07	4.2			
181	2005	3	26	0:56	4.1			
182	2005	6	7	21:32	4.3			
183	2005	6	15	2:19	4.1			
184	2005	8	14	6:33	4.2	15	17.1310	73.8580
185	2005	8	30	8:53	4.8	15	17.1940	73.8450
186	2005	11	13	3:56	4.2	6.7	17.1510	73.7810
187	2005	11	20	18:50	4.0	6.5	17.2210	73.7730
188	2005	12	26	10:46	4.2	4.8	17.1490	73.7860
189	2006	4	17	16:40	4.6	4.2	17.1450	73.7850
190	2007	8	20	19:16	4.2	5.1	17.1450	73.7840
191	2007	11	24	10:57	4.5	5.8	17.1110	73.7050
192	2007	11	24	11:35	4.2	5.5	17.1070	73.7100
193	2008	7	29	19:11	4.0	13.5	17.3240	73.7470
194	2008	9	16	21:47	4.7	9	17.2970	73.7710
195	2009	11	14	13:34	4.7	3.9	17.1170	73.7770
196	2009	12	12	11:51	4.1	4.7	17.1290	73.7830
197	2009	12	12	21:56	4.8	1.5	17.1450	73.7880
198	2009	12	23	3:49	4.1	4.1	17.1230	73.7830
199	2010	1	12	9:25	4.0	4.3	17.1230	73.7850
200	2010	5	24	8:00	4.0	4	17.1280	73.7790
201	2012	4	14	5:27	4.6	8	17.3240	73.7450
202	2013	1	26	2:24	4.1	7.2	17.1520	73.7520
203	2013	9	5	16:40	4.0	5.9	17.2440	73.7230
204	2017	6	3	18:14	4.0	4.0	17.047	73.837
205	2019	6	20	02:17	4.0	4.7	17.121	73.794
206	2021	10	24	17:07	4.1	5.0	17.23	73.69
207	2021	11	15	02:36	4.0	7.0	17.27	73.78

7.16 Koyna Region Is Critically Stressed

In an interesting study, Bansal et al. (2018) reported remote triggering of microearthquakes in the Koyna region following the occurrence of Mw 8.6 Indian Ocean earthquake of 11 April 2012 and its largest aftershock of Mw 8.2. Microearthquake activity started soon after the Love waves generated by these great earthquakes passed through the Koyna region. The largest event of M 3.3 occurred during the first few cycles of Raleigh waves passing through the Koyna region. The increased macroseismic activity lasted for 5 days. There was also an M 4.8 event having occurred 3 days after the occurrence of the earthquake. Bansal et al. (2018) inferred the increase seismicity in the Koyna region following the passing of surface Love and Raleigh waves as an indication that the region is stressed close to critical and is sensitive to remote dynamic triggering.

An interesting study by Madugula et al. (2021) provided another proof of the Koyna region being critically stressed. They observed that every year earthquakes in the Koyna region get triggered following the filling of the reservoir during the monsoon period. Later, earthquakes are triggered

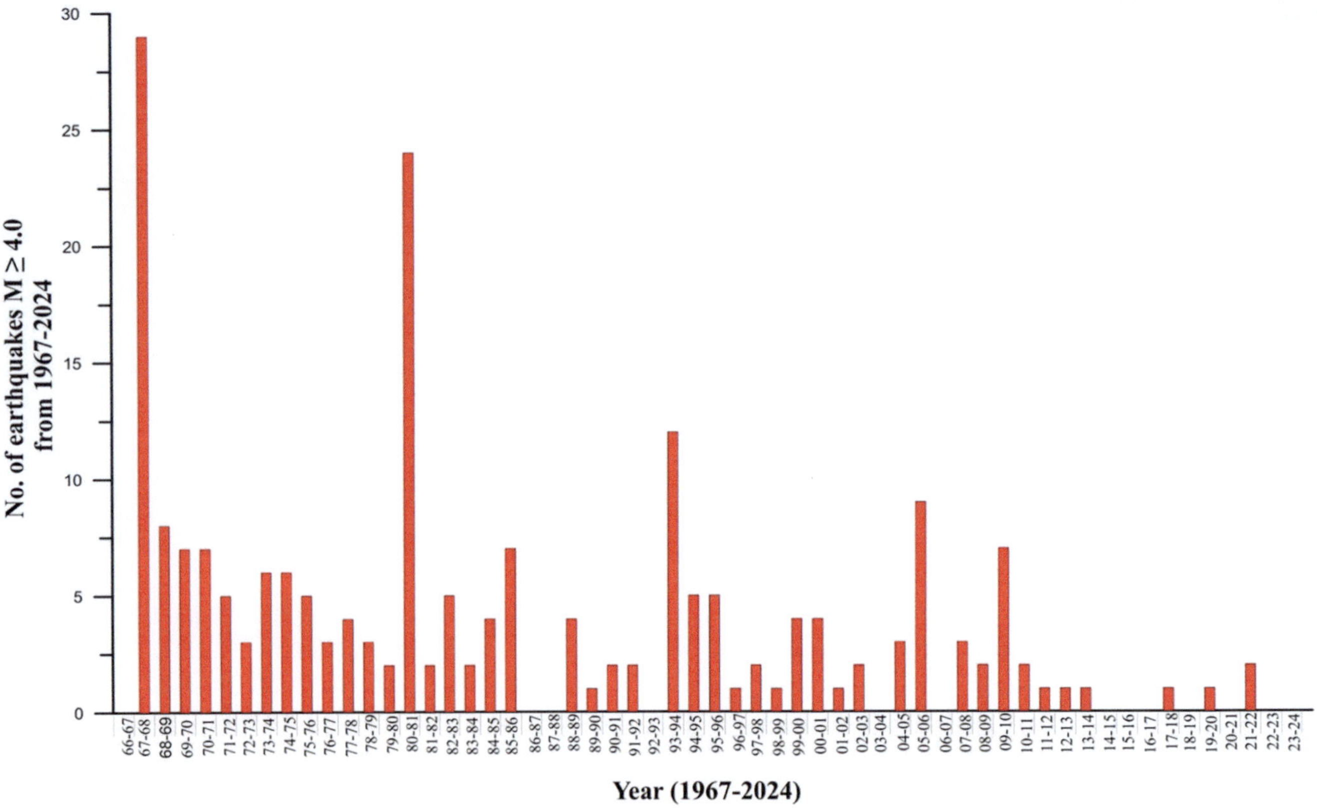

Fig. 7.19 Annual number of earthquakes of M $\geq$ 4.0 during 1967–2024 in Koyna region. Number of M $\geq$ 4.0 earthquakes for annual loading and unloading cycles are shown. For example, there were 29 M $\geq$ 4.0 earthquakes during 1st June 1967–31st May 1968

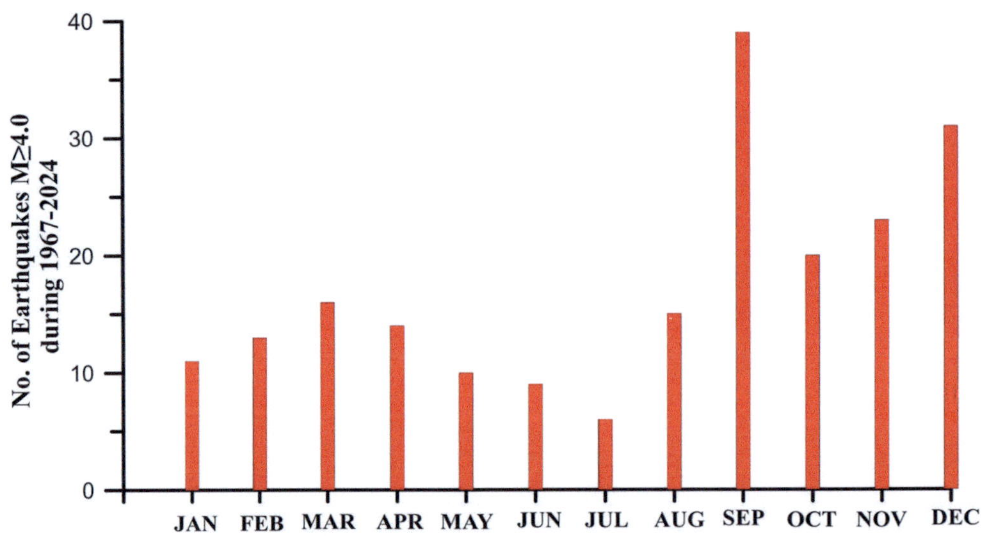

Fig. 7.20 Monthly number of M $\geq$ 4 earthquakes in the Koyna region for the period 1967 through 2024. There is a spurt in seismic activity soon after loading of the reservoirs (end of August–September). Another peak appears in December. The third peak is associated with fast draining of the reservoirs in March

during the reservoir unloading months from January to May. Madugula et al. (2021) found that in the third week of March 2015, the unloading rate at Koyna increased from 0.053 to 0.170 m/day. At the same time the unloading rate increased from 0.065 to 0.170 m/day at the Warna reservoir. This increase in unloading rate was accompanied with the increase in ML $\geq$ 0.8 from 10 to 25. Similar observations of the connectivity in rate of unloading and the number of ML $\geq$ 0.8 was found in the years 2016, 2017 and 2018 (Fig. 7.24). The drilling of the 3 km deep Pilot Borehole in the Koyna region started on the 16 December 2016. Primarily the air-hammer technique was used in drilling. The hammer used weighed 600 kg. It was observed that the drilling episodes increased seismic events around the Pilot Borehole (Madugula et al. 2021). Figure 7.25 depicts the time history of drilling. During the drilling of the Pilot Borehole, there were episodes of rapid

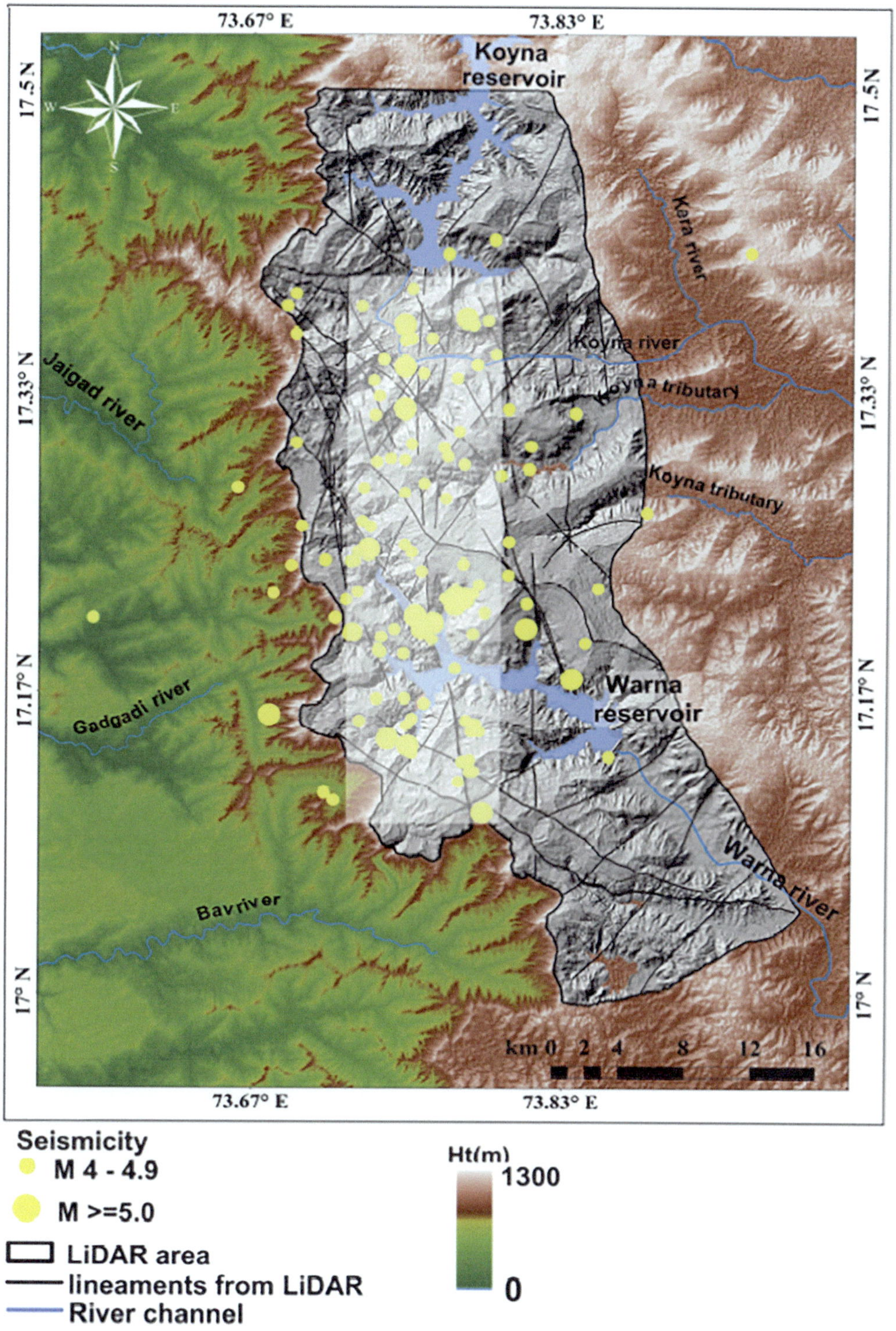

Fig. 7.21 Epicentral distribution of $M \geq 4.0$ earthquakes for the period 1968–2016. A broad 10 km wide N–S zone is demarcated between Koyna and Warna reservoirs along which most of the earthquake occur. The zone shows LiDAR derived Digital Elevation Model (DEM) and lineaments in three directions but N–S fractures controls seismic activity

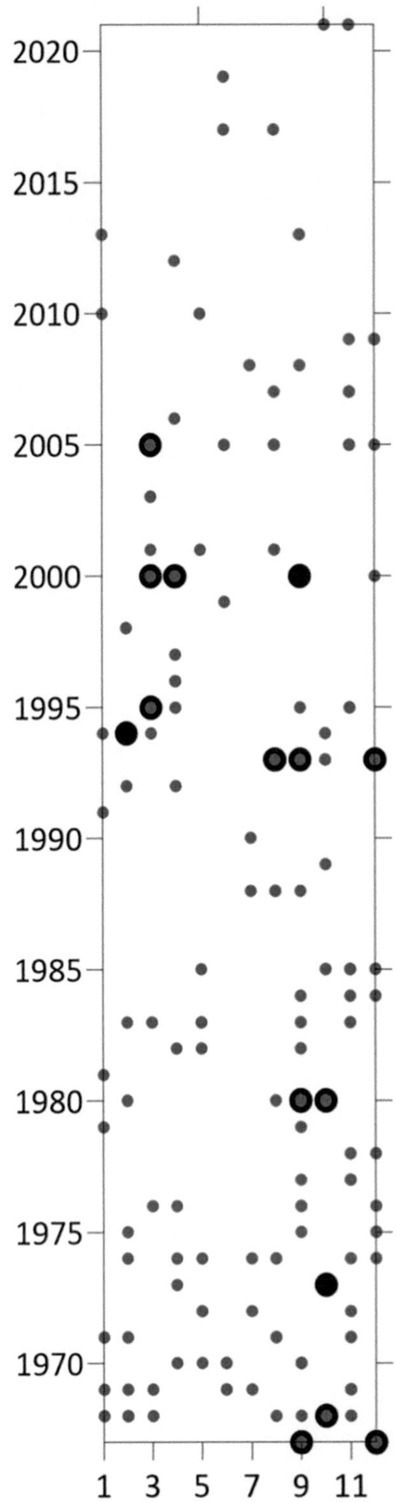

Fig. 7.22 Month-wise occurrence of M ≥ 4 earthquakes in Koyna region for the period 1968 through 2024. Black circles indicate earthquakes of M ≥ 5, while grey small circles are earthquakes of M ≥ 4. Till 1985, most M ≥ 5 earthquakes occurred following the peak reservoir levels in the months of September–December. There was no M ≥ 5 earthquake after 1980 till 1993. The Warna reservoir was impounded in 1993. This gave an impetus to seismic activity as well a shift in M ≥ 5 earthquakes occurring following the emptying of reservoirs during February–April. Figure updated from Arora et al. (2017)

Table 7.7 Frequency of earthquake occurrence in different magnitude ranges in the Warna region from August 2005 to December 2010

Magnitude range	Number of earthquakes
0.1–0.9	3891
1.0–1.9	3179
2.0–2.9	547
3.0–3.9	68
4.0–4.9	15
5.0–5.9	2

drilling lasting for a week or so, followed by a pause for collecting the cores or other drilling requirements. Madugula et al. (2021) chose six rapid drilling episodes which are indicated in Fig. 7.25 by red arrows. These episodes occurred at depths ranging from 1400 to 3000 m, as indicated in Fig. 7.25. It is noted that during the period 13 March–30 June, while in 2015 and 2016, there were 138 and 130 ML $\geq$ 0.8 tremors, this number increased to 323 during this period in the 2017, when rapid hammer drilling was in progress for the Pilot Borehole. The number of ML $\geq$ 0.8 earthquakes during 13 March–30 June again dropped to 186 in the year 2018. Table 7.9 provides the relevant details of the number of ML $\geq$ 0.8 tremors in the Koyna region as well as the rates of unloading of the Koyna and Warna reservoirs and their influence of the increase of seismicity.

The Koyna-Warna region is very well covered with seismometers, including 6 borehole seismometers for a precise and accurate monitoring of tremors of ML $\geq$ 0.8 during the period of drilling of the Pilot Borehole. Madugula et al. (2021) have also analyzed the seismicity observed within 10 km radius of the Pilot Borehole. They report that during 13 March through 30 June 2015, and 2016, there were 68 and 75 ML $\geq$ 0.8 earthquakes within a radius of 10 km of the Pilot Borehole. This number increased to 116 during 13 March–30 June

Table 7.8 List of b values associated with the M $\geq$ 4.0 earthquakes in the Warna region

S. No.	Date of earthquake occurrence	Magnitude (ML)	Associated b value
1	14 August 2005	4.2	0.60
2	30 August 2005	4.8	0.60
3	13 November 2005	4.0	0.94
4	20 November 2005	4.1	0.81
5	26 December 2005	4.2	0.62
6	17 April 2006	4.7	1.30
7	21 May 2006	4.2	1.34
8	20 August 2007	4.3	0.60
9	24 November 2007	4.8	0.64
10	24 November 2007	4.6	0.60
11	24 November 2007	4.1	0.60
12	14 November 2009	5.0	0.60
13	14 November 2009	4.0	0.65
14	12 December 2009	5.1	0.68
15	12 December 2009	4.0	0.60
16	23 December 2009	4.0	0.70

in 2017, the year drilling was in progress. It again dropped to 71 during 13 March–30 June 2018 (Table 7.10).

The remote triggering of earthquakes in the Koyna region following the passage of Love and Raleigh waves of great Mw 8.6 earthquake on 11 April 2012 that occurred in the Indian Ocean supports the view that the Koyna region is critically stressed and a bit of perturbation in the stress field triggers earthquakes. This view is further strengthened by the observation of the influence of the increase in the rate of unloading of the reservoirs in the observed increase in trigger events in the vicinity of the Koyna and Warna reservoirs. An additional support to this view is provided by Madugula et al. (2021), who have demonstrated that hammer drilling for the 3 km

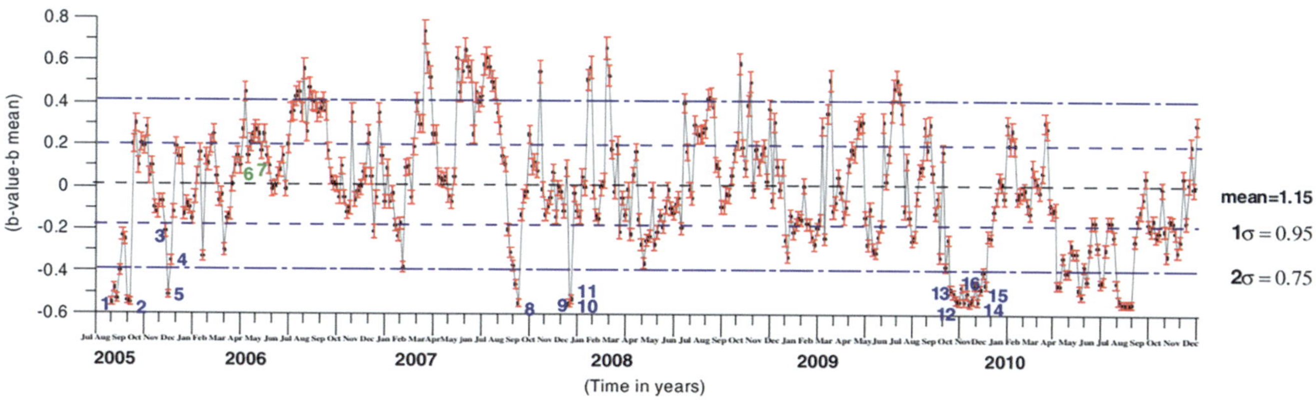

Fig. 7.23 Temporal variation of b values in the Warna region during the period August 2005–December 2010 using a moving average window of 1 month shifted by steps of 5 days. The error bars for each value are also indicated. The numbers indicate the M $\geq$ 4.0 earthquakes listed in Table 7.8. The mean b value and the standard deviation (1σ and 2σ ranges) are represented as dashed lines

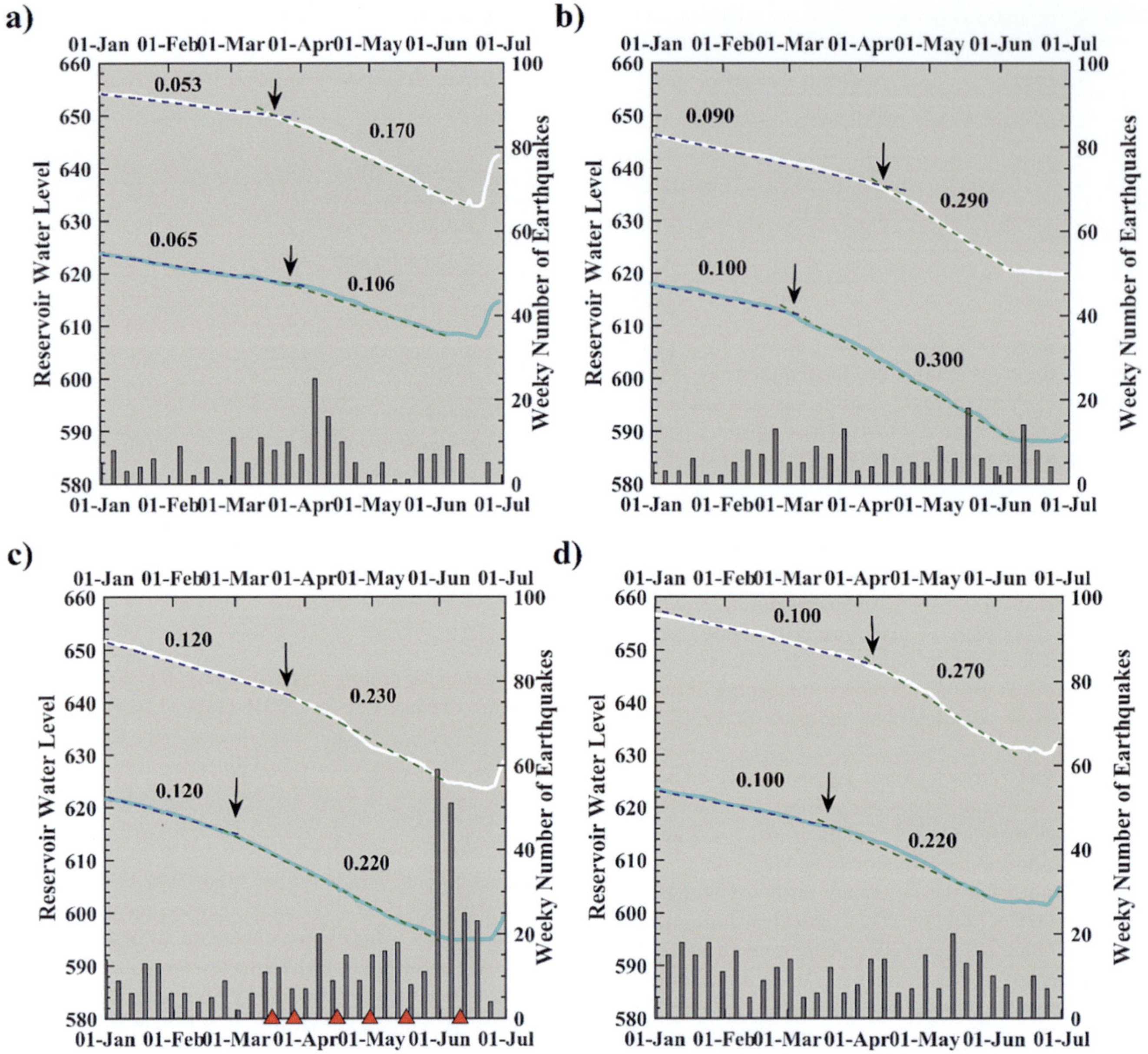

Fig. 7.24 Number of weekly earthquakes with ML $\geq$ 0.8 observed during January–June of **a** 2015, **b** 2016, **c** 2017, and **d** 2018 along with the Koyna (white line) and Warna (cyan line) reservoir water levels. The numbers shown adjacent to reservoir water levels denote the rate of unloading. Black arrows denote the approximate time of change in unloading rate. The red triangles on the bottom x-axis of **c** depict the time of rapid drilling episodes. Blue and green dashed lines depict the approximated linear trends with different unloading rates

deep Pilot Borehole also increased the occurrence of ML $\geq$ 0.8 events within 10 km radius of the Piot Borehole site.

7.17 Earthquake Focal Mechanism

The earthquake focal mechanism in Koyna region has been investigated by several workers (for example Chandra 1976; Gupta et al. 1980, Langston 1976; Yadav et al. 2013; Rao and Shashidhar 2017 and several others). It is observed that basically strike-slip and normal earthquake focal mechanisms dominate the region, while the prevalent tectonics of the region is of compressive nature (Rao and Shashidhar 2017). In an interesting study, way back in 1980, Gupta et al. (1980) observed that earthquakes occurring north of 17° 15′ latitude are basically strike-slip fault dominated, while on the south of 17° 15′, normal fault focal mechanism is prevalent. Figure 7.26 depicts composite focal mechanism solution for 9 M $\geq$ 4 earthquakes after Gupta et al. (1980). The Koyna river flows basically in the north–south direction up to 17° 20′ N and then takes a sharp turn to the east (Fig. 7.26). These observations led Gupta et al. (1980) to infer that the tectonics of the Koyna region is dominated by a set of conjugate faults, with a strike-slip and normal faulting sense of motion.

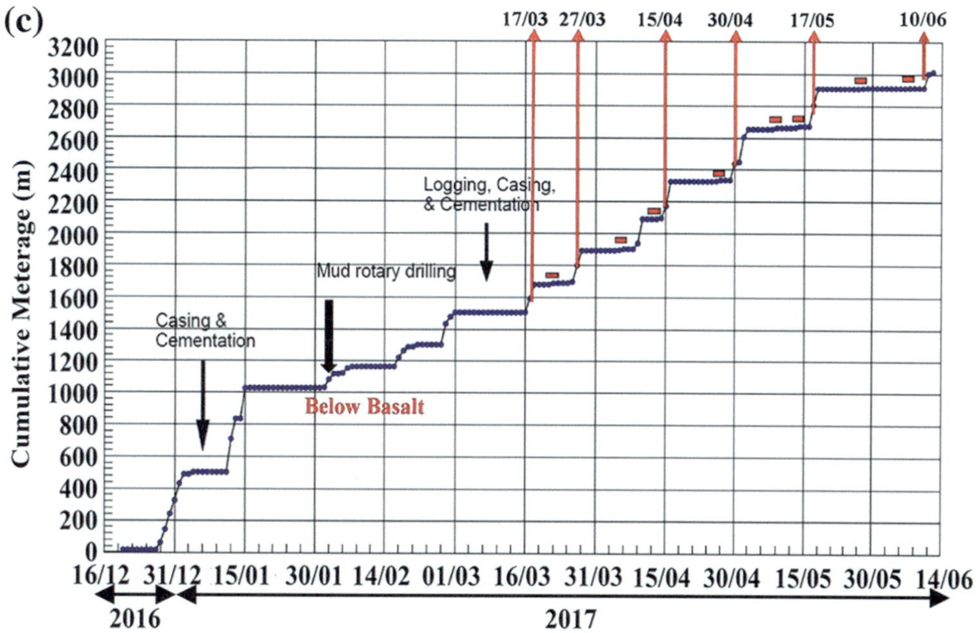

Fig. 7.25 Time history of drilling. The red arrows indicate the six rapid drilling episodes lasting 3–7 days. Date given is the middle of the episode (courtesy Borehole Geophysics Research Laboratory, Karad, India)

Table 7.9 Number of earthquakes observed during 13 March–30 June of 2015, 2016, 2017 (rapid drilling period) and 2018 along with the change in rate of unloading (m/day) observed at Koyna and Warna reservoirs during the unloading period (January–June) of 2015, 2016, 2017 and 2018

Year	Number of earthquakes $M_L \geq 0.8$	Rate of reservoir unloading m/day (Approximate date of change of rate of unloading)	
		Koyna	Warna
2015	138 (13 March–30 June)	0.053–0.170 (20-03-2015)	0.065–0.106 (26-03-2015)
2016	130 (13 March–30 June)	0.090–0.290 (08-04-2016)	0.100–0.300 (01-03-2016)
2017	323 (13 March–30 June)	0.120–0.230 (22-03-2017)	0.120–0.220 (02-03-2017)
2018	186 (13 March–30 June)	0.100–0.270 (06-04-2018)	0.100–0.220 (14-03-2018)

The date around which the rate of unloading changed is mentioned below the unloading rates

Table 7.10 Number of ML ≥ 0.8 events within 10 km radius of the pilot borehole. The seismic activity during 13 March-30 June 2017 is much higher compared to the previous and the following years

Period	Magnitude ($M_L \geq 0.8$)
13 March–30 June 2015	68
13 March–30 June 2016	75
13 March–30 June 2017	**116**
13 March–30 June 2018	71

Bold value indicates the seismic activity during 13 March–30 June 2017 is very high

In a detailed study Rao and Shashidhar (2017) reported focal mechanism solutions of 50 earthquakes of M ≥ 4 that occurred during 1967 through 2017 in the Koyna region. The epicenters of these earthquakes and seismic stations in operation in the Koyna region are shown in Fig. 7.27. The focal mechanisms obtained are shown in Fig. 7.28. They observed that from 1967 through 1973, strike-slip faulting dominated. From 1973 through 1989, normal faulting dominated. Again 1989 through 1995, it was strike-slip faulting dominating, followed by normal faulting during 1995 through 2011. There was again a switch over to strike slip faulting for the 2012 M 4.8 earthquake. It is noted that north of 17° 15′ N latitude, strike-slip focal mechanism dominates, while south of 17° 15′ N latitude normal faulting prevails.

The major conclusions drawn by this detailed study of focal mechanisms of Koyna earthquakes by Rao and Shashidhar include:

1. The Koyna region is characterized by strike-slip and normal faulting in the Indian Shield region which is dominated by thrust faulting.

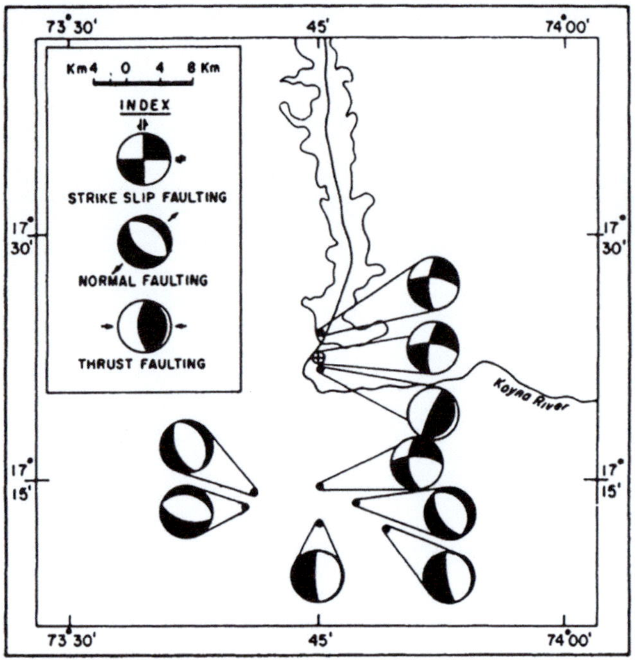

Fig. 7.26 Composite focal mechanism solution for nine M > 4 Koyna earthquakes (adapted from Gupta et al. 1980). Focal mechanism for the December 10, 1967 earthquake (+) is from Singh et al. (1975). Earthquakes south of 17° 15′ latitude have normal fault mechanism

2. Focal mechanism studies in the Koyna region have advanced from the first motion polarity approach to whole waveform inversion method.
3. Precise depths estimated from whole waveform inversion have shown that the focal depths extend to 13 km in the Koyna region, while they are typically 4–6 km in Warna region.
4. The Donachiwada Fault has been well delineated through precise modelling of the 14 April 2012 M 4.8 earthquake.
5. A periodic variation of the stress field in the Koyna region alternating between strike-slip and normal faulting has been inferred.

7.18 Overview of Deccan Basalts in the Koyna Region

An excellent overview is provided by Subbarao and Courtillot (2017) on the basalt in and around Koyna region. They have reviewed the earlier chemo- and magnetostratigraphic studies along with the radiometric dates. The significance of these observations is highlighted and their implication in understanding the present-day reservoir triggered seismicity in the Koyna region is underlined.

There has been a significant contribution in establishing the detailed stratigraphy of Western Ghats which consist of ~ 3000 m of lava pile. This pile has been divided into three subgroups, namely, the Kalsubai, the Lonavala, and the Wai. These are divided in 12 well defined Formations.

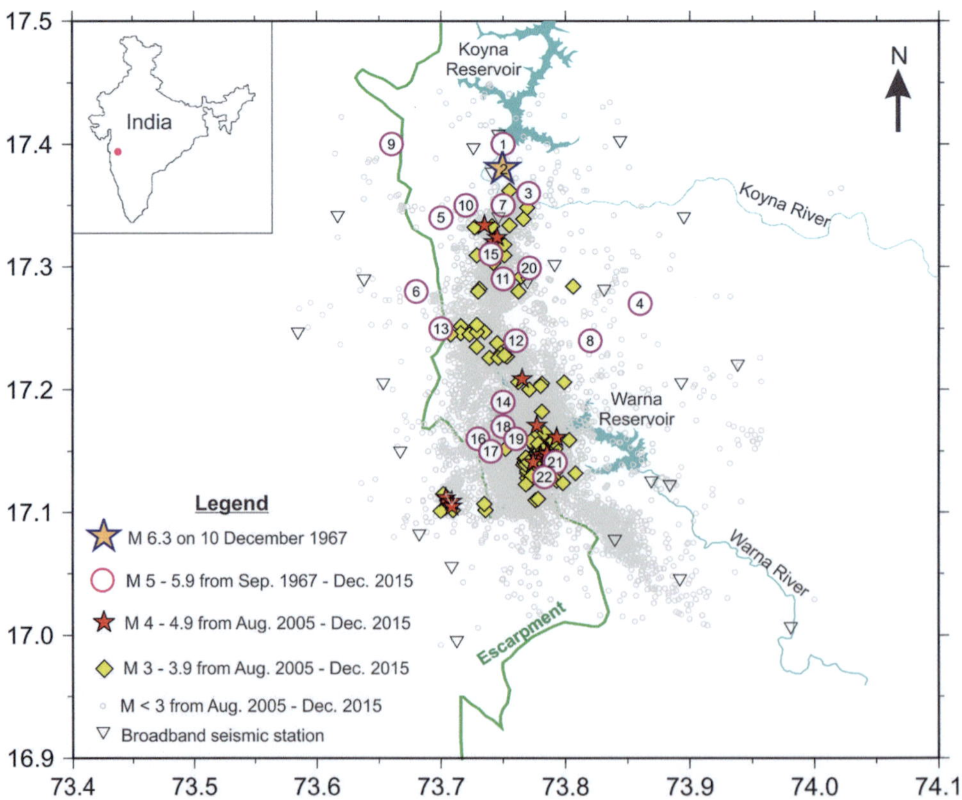

Fig. 7.27 Seismic station network of CSIR-NGRI in the Koyna–Warna region, along with seismicity distribution comprising earthquakes of M ≥ 5 since 1967 and smaller ones during 2005–2015. Inset: Map of India indicating the Koyna–Warna region

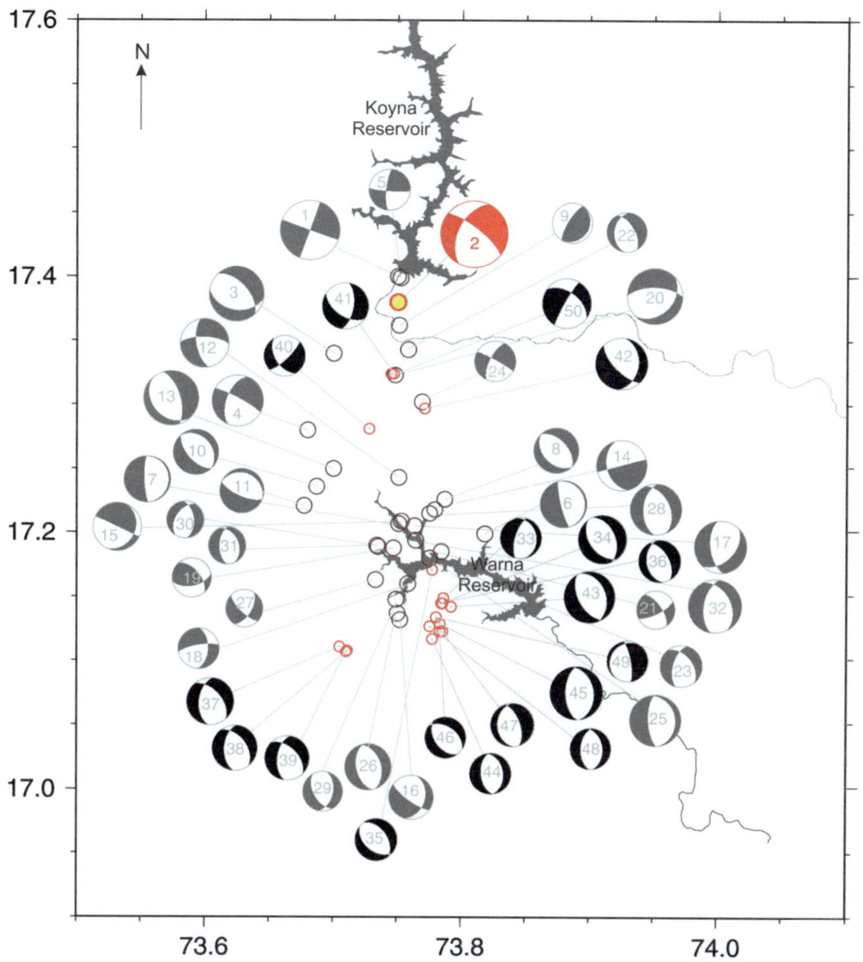

Fig. 7.28 The comprehensive set of 50 earthquake focal mechanism solutions in the Koyna–Warna region compiled from previous studies (gray) and from moment tensor inversion of waveform data by the authors in their previous work and the present study (black). #2 is for the focal mechanism for the 10 December 1967, main earthquake. The focal mechanism after 2005 are indicated in red circles

Figure 7.29a, b show the extent of Deccan Traps in India and simplified stratigraphic column of Deccan Basalt Group in Western Ghats of India. The Lonavala and Wai Subgroups of flow lie south of Mahabaleshwar, Koyna and Belgaum and in east around Nagpur (Fig. 7.30a–c). This clearly indicates a southward plunging anticlinal structure.

Figure 7.31 shows stratum contours drawn to the base of the Thakurvadi, Bushe, and Mahabaleshwar Formations which basically represent the southernly dipping anticlinal feature, which becomes monoclinal south of 17° 15′ N latitude. The axis of the anticline is parallel to the scarp line of Western Ghats Escarpment.

Summarizing their detailed study, Subbarao and Courtillot (2017) observe that:

1. The Koyna region lies south of Mahabaleshwar Ghat with about 2000 m thick Deccan lava flows, with Mahabaleshwar lying at the edge of the Western Ghats Escarpment. Several multi-disciplinary studies conducted in the past have made it one of the best studied plateaus in the world.
2. Chemo stratigraphic studies place Mahabaleshwar, Koyna and Warna lava sequences into younger Wai group having the presence of Poladpur, Ambenali, Mahabaleshwar, and Desur Formations.
3. In terms of magnetostratigraphic classification, all the flows below Mahabaleshwar Formation are reversely magnetized. Whereas almost 90% of Mahabaleshwar Formation and the younger Desur and Panhala Formations are normally magnetized.
4. The Normal/Reverse magnetization transition is seen up to Belgaum, passing through the Koyna-Warna region. It is important to note that the north to south flowing Koyna River lies parallel to the axis of Mahabaleshwar anticline.
5. Based on the data from boreholes drilled in the region, it is found that the basement depths are within a variation of 200 m on the western part while the variation is up to 500 m on the east side.
6. Using the available data, it is inferred that the Koyna-Warna region comprises of Desur, Mahabaleshwar, Poladpur, Bushe and Khandala formations, covering N/R magnetic polarity transition at Mahabaleshwar Formation that matches with the lowermost part of C29n and upper parts of C29r chrons.
7. Surface geochronological studies place the massive volcanic eruption around 64.7 Ma.

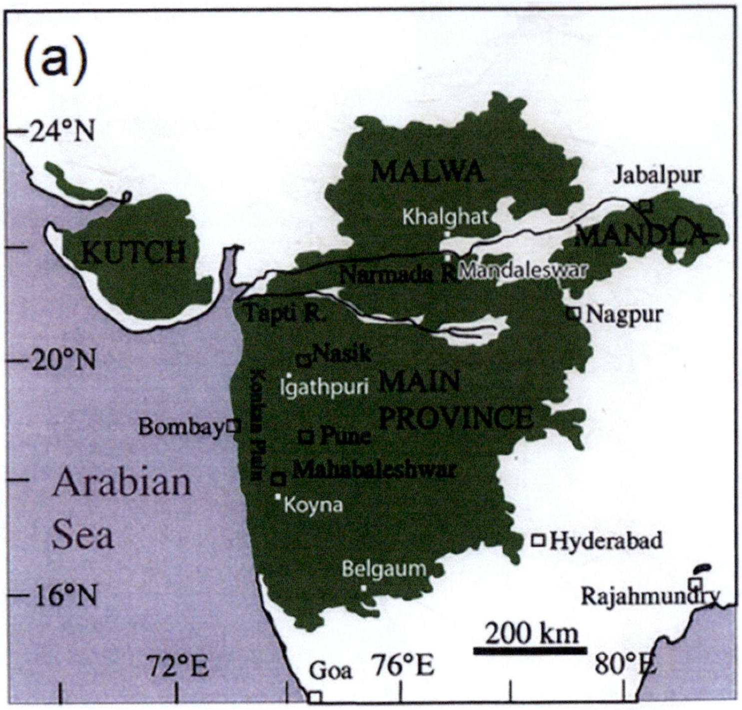

(b)

Group	Subgroup	Formation	Member / Flow	Thickness M	Polarity
DECCAN BASALT GROUP	Wai	Desur			N
		Panhala		>175	
		Mahabaleshwar		280	
		Ambenali		500	R
		Poladpur		375	
	LONAVALA	Bushe		325	
		Khandala		145	
	Kalsubai	Bhimashankar	Monkey Hill GPB Giravili GPB	140	
		Thakurvadi	Manchar GPB	650	
		Neral	Tunnol-5 GPB	100	
		Igatpuri	Kashele GPB	> 500	
		Jawhar	Thalghat GPB	> 200	

Fig. 7.29 **a** The extent of Deccan Traps is shown on the map of India. Important names of localities are shown on the map where samples were collected and/or discussed in this paper. **b** Simplified stratigraphic column of the Deccan Basalt Group in the Western Ghats, showing the position of the giant plagioclase basalts (GPBs) at the top of each Formation in the Kalsubai Subgroup (Beane et al. 1986; Devey and Lightfoot 1986; Subbarao and Hooper 1988; Cox and Hawkesworth 1985). Thickness and magnetic polarity of each formation is based on various mapping programmes referred to in the text

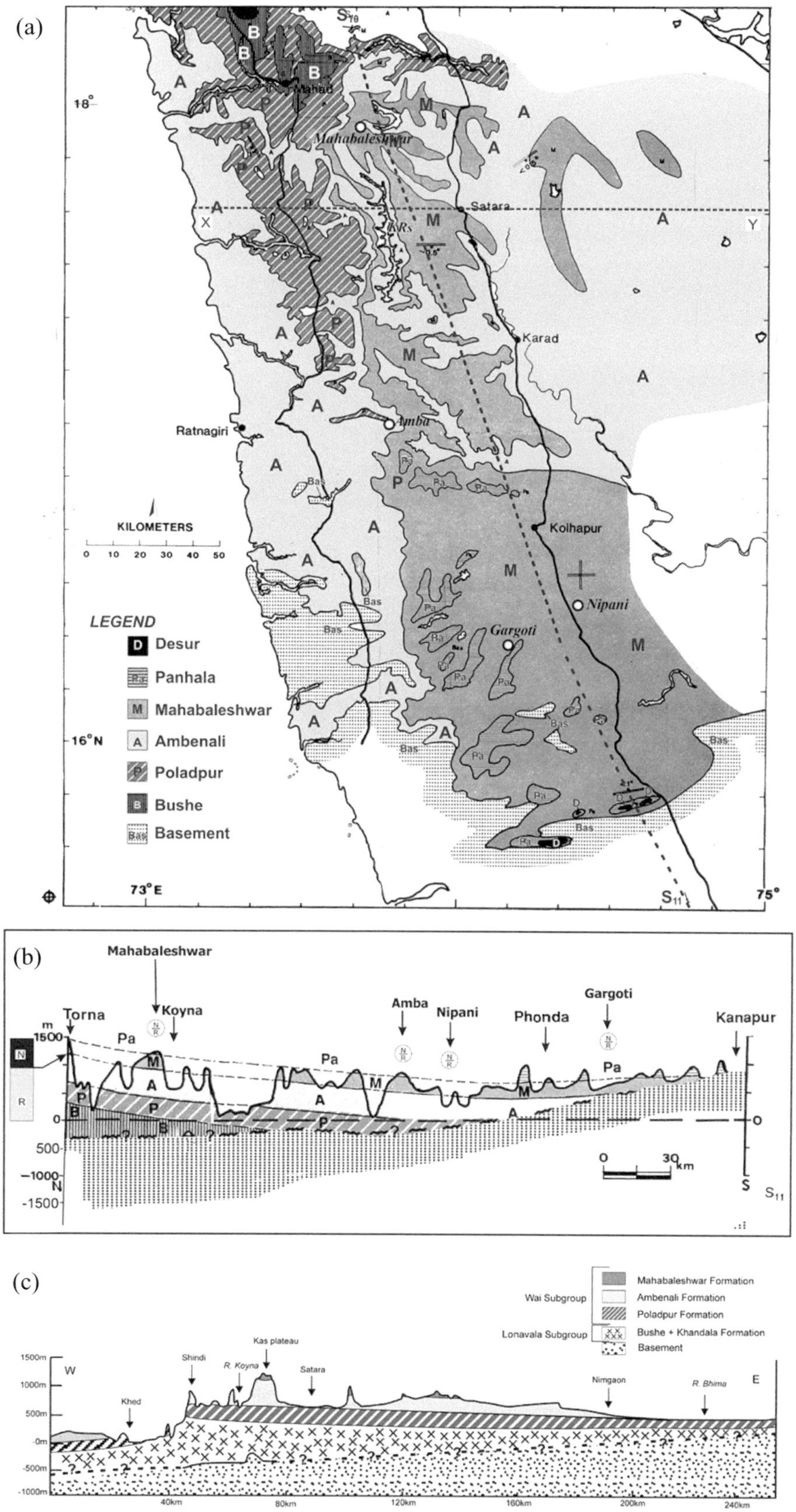

Fig. 7.30 **a** Geological sketch map of the southern part of the Deccan Volcanic Province (Subbarao and Hooper 1988; Devey and Lightfoot 1986; Beane et al. 1986; Mitchell and Cox 1988). **b** N–S section (see **a**) for traverse line showing over stepping of flows from N to S (see **a** for legend) (Subbarao and Hooper 1988). Important localities where reversed to normal polarity changes are reported. N/R is shown on the polarity log on the left side of the section where the change occurred from Mahabaleshwar upwards (this study). **c** E–W section along Satara and Khed (see **a** for geological set up). This section lies south of Mahabaleshwar and continues to show younger Wai Subgroup flows followed by Lonavala Subgroup which is resting over the Archean basement. The solid line with a small hump marked on the E–W basement line covers seven boreholes passing through the basement are also shown

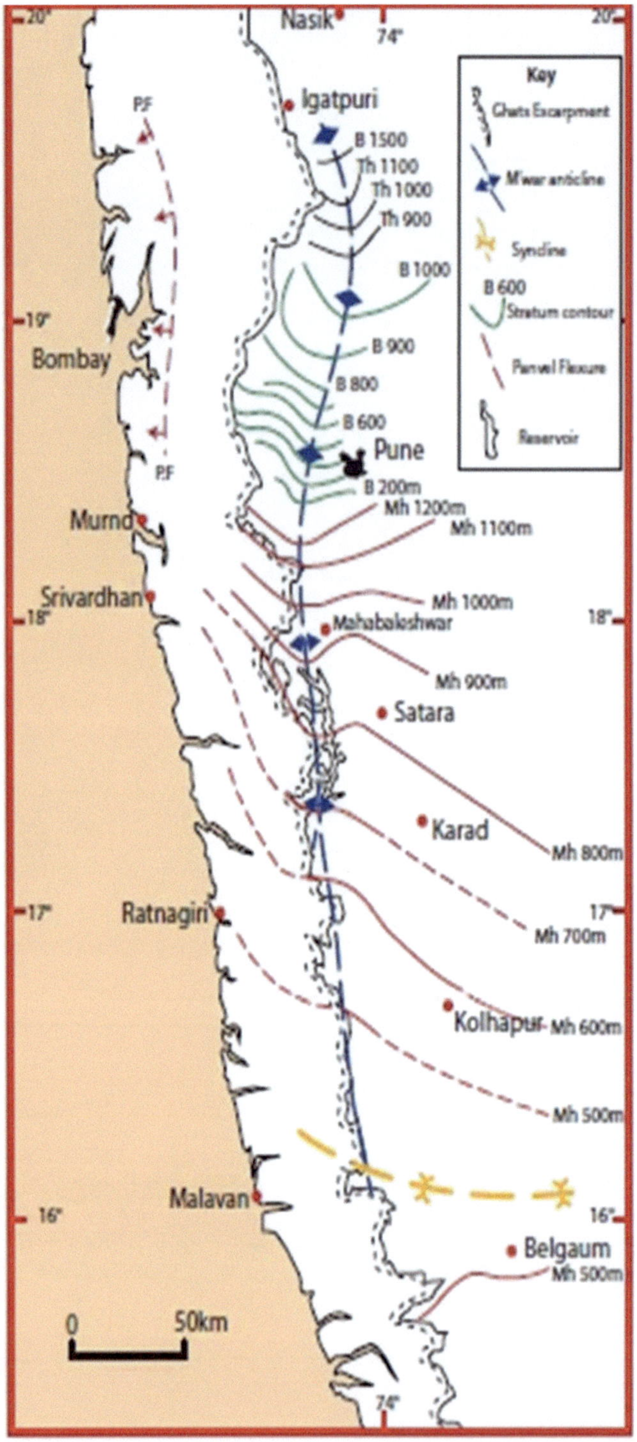

Fig. 7.31 Stratum contours drawn to the base of the Thakurvadi, Bushe and Mahabaleshwar Formations which together define the large southerly dipping anticlinal feature which becomes essentially monoclonal south of 17° 15′ north. It is important to note that the axis of the anticline is parallel with the scarp line of the Western Ghats Escarpment. The E–W trending synclinal trough and associate dip reversal in the southern most parts is an important feature to be noted (Devey and Lightfoot 1986; Khadri et al. 1988; Subbarao and Hooper 1988; Bodas et al. 1988; Widdowson and Mitchell 1999)

7.19 Permeability Modelling for RTS at Koyna

Permeability of rocks play an important role in pore-pressure diffusion that is responsible for creating stress perturbation resulting in RTS. In an interesting study, Hazarika et al. (2017), basing on poroelastic theory, have applied 2-D finite element modelling to study evaluation of pore pressure after 5 years of the impoundment of the Shivaji Sagar Lake formed by the Koyna Dam. The reservoir impoundment took place in 1962, and the M 6.3 Koyna earthquake occurred in 1967. They used permeability of 10^{-16}–10^{-14} m^2, which are the available values of permeability of granitic rocks. A permeability of 10^{-16} m^2 does not produce any significant change in pore pressures at the focal depths of 8 km (the estimated depth of the Koyna earthquake of 10 December 1967). If one increases the permeability by tenfold, a significant change after 5 years is seen at 2 km depth only. The permeability of granites as measured in laboratories is of the order of 10^{-16} m^2. According to Hazarika et al. (2017), a permeability of 10^{-14} m^2 is required to cause an adequate change pore fluid pressure in the range of 0.1–1.0 MPa to a depth of ~ 10 km. However, the cores collected from the granitic basement at depths of ~ 1500 m provide evidence of existing fractures that facilitates water channelization that ultimately leads to the changes in pore pressure facilitating earthquake occurrence.

7.20 Waxing and Waning of RTS

In an interesting study of the spread of RTS in the Koyna region, Shashidhar et al. (2017) have shown that RTS is widely spread around the Koyna and Warna reservoirs during the loading phase of the reservoirs, however, it gets restricted to the immediate vicinity of the reservoirs during the unloading phase. They utilized the data from 23 broadband surface seismographs and 6 borehole seismographs for the period January 2016 through May 2017. During this period 2487 tremors of M_L – 0.8 to 3.7 were recorded. A close association of RTS with the loading and unloading of the Koyna and Warna reservoirs is observed (Fig. 7.32). In Fig. 7.33, it may be noted that during January–May 2016 unloading phase of the Koyna and Warna reservoirs, the epicenters were clustered close to Donachiwada fault and the periphery of the Warna reservoir, the situation changed during loading phase from June to December 2016, as the epicenters spread out. However, during the unloading phase from January to May 2017, the epicenters again got concentrated in the vicinity of the Donachiwada fault and the periphery of the Warna reservoir. This waxing and waning of the RTS in the vicinity of the Donachiwada fault and the periphery of the Warna reservoir is inferred in terms of the effect of the reservoir water levels and its influence on pore fluid pressure which contribute to triggering earthquakes.

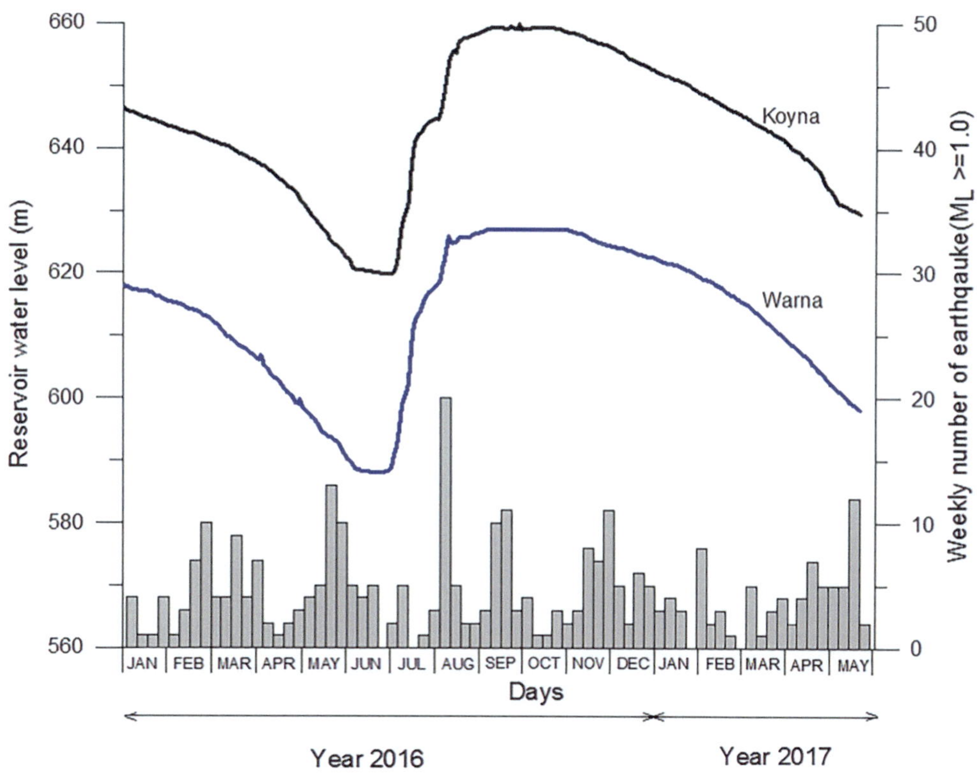

Fig. 7.32 Reservoir water levels at Koyna and Warna reservoirs and the weekly number of ML ≥ 1.0 earthquakes during January 2016–15th May 2017

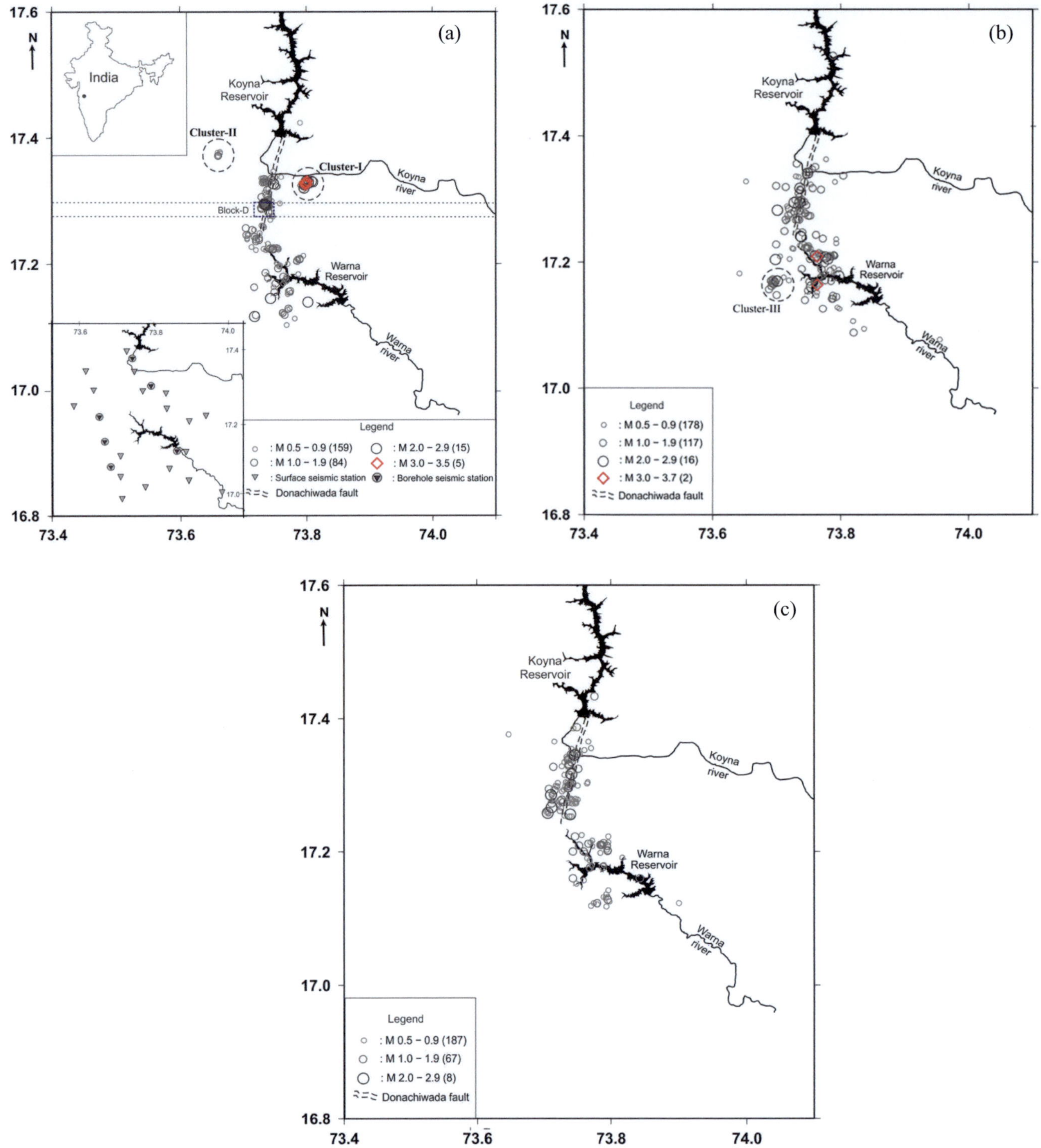

Fig. 7.33 Seismicity of the Koyna-Warna region during **a** January–May 2016. Dotted circles encompass the new clusters. The dotted lines indicate the swath of 2 km area of block D. In the inset: (top left) India map indicates the study region; (bottom left) seismic stations. **b** June–December 2016. Dotted circle encompass the new cluster. **c** January–15th May 2017. It may be noted that in **a** and **c** earthquakes are close to Donachiwada fault and widespread in **b**

Table 7.11 Koyna water levels and derived statistics (1961–2016)

1	2	3	4	5	6	7	8	9	10	11
Serial no. of M ≥ 5 events	Date of occurrence of M ≥ 5 earthquakes	Date when water level reached minimum	Minimum water level (m) (LWL)	Date when water level reached maximum	Peak water level (m) (P1)	Filling days	Level rise (m)	Duration required for water level to P1-5 m	Maximum filling rate (m/week)	Difference between P1 and previous maximum
		07.05.61	586.7	11.07.61	624.7	56	39.0	76	14.1	
		04.07.62	615.5	15.08.62	636.7	43	21.2	102	17.3	12.0
		28.06.63	617.5	14.08.63	654.1	45	36.6	88	13.8	17.4
		26.06.64	635.2	13.08.64	654.9	49	19.7	74	8.0	0.8
		18.06.65	635.3	22.07.65	655.2	35	19.9	60	10.0	0.3
		20.06.66	630.5	03.08.66	652.7	45	22.3	101	9.8	– 2.5
		29.06.67	620.7	04.10.67	656.9	98	36.3	126	12.6	1.7
1	13.09.67 (M 5.2)									
2	10.12.67 (M 6.3)									
3	10.12.67 (M 5.0)									
4	11.12.67 (M 5.2)									
5	12.12.67 (M 5.4)									
6	12.12.67 (M 5.0)									
7	12.12.67 (M 5.0)									
8	24.12.67 (M 5.5)									
9	25.12.67 (M 5.1)									
10	25.10.68 (M 5.4)									
		04.07.68	616.8	29.08.68	646.7	57	30.6	93	9.3	– 10.2
		19.06.69	614.7	08.08.69	653.7	51	39.0	80	13.2	– 3.2
		14.06.70	620.8	21.08.70	653.6	69	32.8	77	8.1	– 3.3
		04.06.71	620.3	01.09.71	653.3	88	33.1	68	10.5	– 3.6
		26.06.72	612.7	28.08.72	645.8	64	32.8	66	14.4	– 11.1
11	17.10.73 (M 5.2)	10.06.73	610.6	27.09.73	657.9	79	47.3	137	22.6	+ 1.0
		03.07.74	614.8	02.09.74	649.6	62	34.8	98	12.0	– 8.3
		20.06.75	611.6	07.09.75	657.9	80	46.3	106	9.6	0.0
		03.06.76	631.8	05.09.76	657.9	95	26.1	94	7.0	0.0
		24.06.77	626.4	31.08.77	657.9	69	31.5	97	9.6	0.0
		16.06.78	621.2	25.08.78	657.9	71	36.7	87	9.6	0.0
		25.06.79	610.9	19.08.79	656.5	56	45.6	97	14.4	– 1.4
12	02.09.80 (M 5.3)	22.06.80	621.2	03.09.80	658.1	74	36.9	98	13.9	+ 1.2
13	20.09.80 (M 5.3)									
		26.06.81	620.7	01.10.81	657.8	98	37.0	72	12.0	– 0.3
		19.06.82	625.5	01.09.82	657.5	75	32.0	90	8.0	– 0.6

(continued)

Table 7.11 (continued)

1	2	3	4	5	6	7	8	9	10	11
Serial no. of M ≥ 5 events	Date of occurrence of M ≥ 5 earthquakes	Date when water level reached minimum	Minimum water level (m) (LWL)	Date when water level reached maximum	Peak water level (m) (P1)	Filling days	Level rise (m)	Duration required for water level to P1-5 m	Maximum filling rate (m/week)	Difference between P1 and previous maximum
		20.06.83	625.0	27.09.83	657.9	100	32.9	86	7.5	– 0.2
		13.06.84	628.0	03.09.84	657.5	83	29.5	95	6.2	– 0.6
		16.06.85	628.7	07.09.85	656.4	84	27.8	100	8.0	– 1.7
		18.06.86	631.0	26.08.86	656.8	70	25.8	103	6.0	– 1.3
		30.06.87	618.5	23.10.87	650.7	116	32.2	94	11.9	– 7.4
		04.07.88	623.6	26.09.88	658.0	85	34.4	85	14.4	– 0.1
		15.06.89	629.2	04.10.89	654.2	112	25.0	71	10.0	– 3.9
		17.06.90	623.9	30.08.90	658.1	75	34.2	126	8.0	0.0
		06.06.91	630.4	29.08.91	657.9	85	27.5	75	7.2	– 0.2
		09.07.92	614.1	07.09.92	658.0	61	43.9	77	13.2	– 0.1
14	08.12.93 (M 5.2)	15.06.93	621.9	06.09.93	658.0	85	36.2	140	9.4	– 0.1
15	01.02.94 (M 5.4)	13.06.94	623.6	02.09.94	658.2	82	34.6	70	8.0	+ 0.1
		22.06.95	615.2	15.09.95	652.0	85	36.9	97	13.8	– 6.2
		16.06.96	618.4	07.10.96	655.3	111	37.0	91	12.8	– 2.9
		16.06.97	634.4	23.08.97	658.1	68	23.7	123	7.3	– 0.1
		26.06.98	627.0	03.10.98	657.9	99	31.0	95	9.2	– 0.3
		16.06.99	635.3	05.10.99	657.9	111	22.7	95	7.3	– 0.3
16	12.03.2K (M 5.2)									
17	06.04.2K (M 5.1)									
18	05.09.2K (M 5.3)	04.05.2K	633.35	09.09.2K	655.9	128	22.3	99	10.5	– 2.3
		11.06.01	630.2	01.09.01	655.6	82	25.4	119	8.0	– 2.6
		14.06.02	627.1	11.09.02	658.2	89	31.1	60	7.8	+ 0.0
		18.06.03	628.9	19.09.03	656.9	93	28.0	71	5.9	– 1.3
		14.06.04	628.4	01.10.04	658.2	109	29.8	71	9.1	+ 0.0
19	14.03.05 (M 5.2)	23.06.05	627.6	22.09.05	658.6	91	31.0	58	14.1	+ 0.4
		28.05.06	632.1	19.09.06	659.3	114	27.2	82	13.6	+ 0.7
		22.06.07	631.8	28.09.07	659.4	98	27.6	90	15.0	+ 0.1
20	16.09.08 (M 5.2)	30.06.08	635.3	17.09.08	659.5	79	24.2	136	9.1	+ 0.1
21	14.11.09 (M 5.0)	02.07.09	634.2	08.10.09	659.4	98	25.2	121	9.4	– 0.1
22	12.12.09 (M 5.1)									
		14.06.10	633.8	26.09.10	659.4	104	25.6	132	10.8	– 0.1
		13.06.11	634.1	29.08.11	659.2	77	25.1	77	7.8	– 0.3
		03.07.12	630.9	27.10.12	659.6	116	28.7	61	8.5	+ 0.1
		07.06.13	633.7	05.10.13	659.5	120	25.8	58	5.7	– 0.1
		11.07.14	620.2	07.09.14	659.3	58	39.1	110	13.9	– 0.3
		19.06.15	632.7	30.08.15	652.8	72	20.1	101	9.2	– 6.8
		23.06.16	619.8	23.09.16	659.9	92	40.1	73	12.1	+ 0.3

7.21 Loading of the Koyna Reservoir Parameters and RTS

The correspondence between factors like rate of loading, highest water levels reached, duration of retention of the high-water levels, and their correspondence with M $\geq$ 5 RTS events has been pointed out by Gupta et al. (1972), Gupta and Rastogi (1974), and Gupta et al. (2002). It was also found that M $\geq$ 5 events invariably occurred when the previous water maxima in the Koyna reservoir was exceeded. Gupta et al. (2002) have pointed out that during 1967 the lake was filled to 656.9 m exceeding the previous maxima by 1.7 m, with the maximum rate of loading being 12.6 m/week. The high-water levels were retained for 126 days, and this was followed by intense seismic activity including the 10 December 1967 M 6.3 earthquake. For the next five years no M $\geq$ 5 earthquakes occurred. However, in 1973, the previous water maxima were exceeded, and the maximum rate of loading was 22.6 m/week. The subsequent M 5 earthquake took place on 17 October 1973, registering a magnitude of 5.2. Anuradha and Mallika (2017) have extended the study of these parameters till 2017, pointing out that out of 22 M $\geq$ 5 events only 4 occurred during the unloading phase. The first of these occurred in 1994 after the impoundment of the Warna reservoir in 1993. Table 7.11 provides the information on Koyna water levels, M $\geq$ 5 earthquakes and Koyna water levels derived parameters. Anuradha and Mallika (2017) conclude that the rate of loading, high water levels reached and duration of the retention of these high-water levels as well as the fact whether the previous water level maxima has been exceeded or not, continue to influence RTS at Koyna.

7.22 Studies of Cores Obtained from the Boreholes and Basement in the Koyna Region

Several studies have been carried on the cores obtained from the 9 boreholes penetrating the basement at around 1500 m depths as well as the 3000 m deep Pilot Borehole. A few of these are included in the following.

7.23 Magnetic Mineralogical Variability

In an interesting study, Sangode et al. (2017) reported magnetic mineralogical variability in 248 m long core obtained from Koyna borehole # 7 (KBH 07) located at 17° 18′ 07″ N; 73° 47′ 28″ E at an elevation of 960 m above mean sea level. The borehole was drilled to the basement and the cores obtained were sampled at 10 m interval for magnetic mineralogical studies. The studied parameters helped in identifying magma compositional changes, zones of oxidative conditions, and rapid/slow cooling intervals. The rock magnetic properties were discovered to be varying in a complex manner independent of flow units. This has been inferred as reflecting temporal magnitude of compositional variability. These data are helpful in inter-core correlations.

7.24 Granite-Gneiss Basement

Misra et al. (2017) have inferred a granite-gneiss basement in the Koyna region. They studied cores obtained from 4 boreholes, namely KBH-1, KBH-5, KBH-6 and KBH-7. These boreholes are well distributed around the Koyna region with KBH-1 located in the north, KBH-5 in west, and KBH-6, and KBH-7 located in the east. All these 4 boreholes penetrate the entire Deccan Trap and penetrate a few hundred meters into the granitic basement. Table 7.12 provides the locations of these 4 boreholes. Figure 7.34 shows the photomicrographs of the basement granitoids (from Misra et al. 2017). The major findings of this interesting study show that the basement granitoids are made up of mostly granite-gneiss, granite and magmatic gneiss; occurrence of strained quartz and unstrained plagioclase feldspar is inferred in the basement granitoids from study of petrology and microstructures; the basement sections are characterized by localized fault zones, evidenced by fault breccia, fault gauge and slickensides; and, the inferred fractures in the basement provide the pathways for water channelization.

7.25 Forty-Six Lava Flows Identified

In a very interesting study, forty-six lava flows have been identified in the 1500 m deep KBH-7. This borehole is located east of the Donachiwada fault, in Panchgani village (Fig. 7.35). Mishra et al. (2017) have identified these lava flows through mesoscopic observations supported by petrographic studies. The thickness of the Deccan Basalt in Panchagani is 1251.2 m. The 46 lava flows identified include 40 simple flows and 6 compound lava flows. The pile of lava flows lies almost directly over the underlying granitic basement, except a thin layer of 0.9 m infratrappean sediments. The basalt lava pile is found to consist of 74% of massive basalt and 24% vesicular/amygdaloidal basalt. 7 of these flows are separated by red-bole beds. An interesting feature is the discovery of nine giant plagioclase basalt in a single vertical section. Figure 7.36 gives a bar graph showing the thickness of individual lava flows. Flow # 1 is just above the granitic basement.

Mishra et al. (2017) point out that there have been earlier studies of the Deccan Basalt flows since 1950, such as by Auden (1952), Srinivasan and Dutta (1960, 1961); and studies by the Geological Survey of India after the 1967 Koyna M 6.3 earthquake (GSI 1968; Raman et al. 1984). These studies

Table 7.12 List of boreholes of the present study along with locations, total depths and thickness of Deccan Trap

Borehole designation	Location	Total depth w.r.t. borehole collar (m)	Latitude, longitude	Elevation of borehole collar (w.r.t. msl) (m)	Depth to base of Deccan Basalt (w.r.t. borehole collar elevation) (m)
KBH-1	Rasati	1522.5	17° 22′ 38.5″ N 73° 44′ 27.8″ E	580 m	932.50
KBH-5	Phanasavale	906.5	17° 09′ 1.2″ N 73° 40′ 2.6″ E	131 m asl	499.25
KBH-6	Ukhalu	1500.7	17° 07′ 33.1″ N 73° 52′ 08.8″ E	570 m asl	775.04
KBH-7	Panchgani	1503.0	17° 18′ 07″ N 73° 47′ 28.2″ E	960 m asl	1251.23

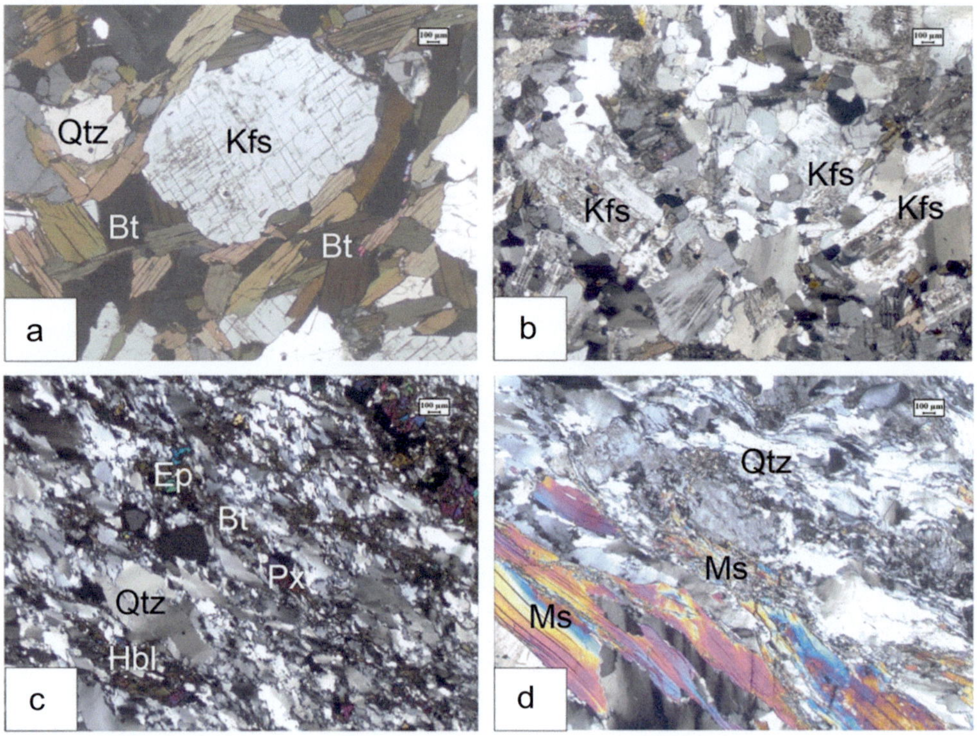

Fig. 7.34 Photomicrographs of basement granitoids from Koyna region showing major mineralogy, textures and deformations. **a** Photomicrograph of granite-gneiss from KBH-5, showing biotite-defined gneissosity and mosaic texture within minerals. Here, biotites (Bt) are wrapped around the large crystal of K-feldspar (Kfs) that indicates metamorphic texture. **b** Photomicrograph of pink granite from KBH-7 showing phenocrysts of K-feldspars with interlocking arrangement indicating igneous texture. **c** Photomicrograph of sheared granite-gneiss in KBH-5 representing highly strained and dynamically recrystalized quartz grains. Biotite (Bt), Hornblende (Hbl) and Pyroxene (Px) retain the gneissosity. **d** Photomicrograph of recrystallized muscovite (Ms) and dynamically recrystallized quartz (Qtz) grains indicating deformation under greenschist facies condition in KBH-5 basement granitoids

provided useful general description of the lava flows in the area. However, this is the first time that the entire lava pile from surface to the basement has been investigated.

7.26 Elastic Properties and Rock Strength of Basement Granitoids

Elastic properties and rock strength of basement granitoids obtained from the Koyna boreholes are reported by Goswami et al. (2017a). KBH-7 is the borehole located east of the Donachiwada fault (Fig. 7.35). This is the first time when granite, granite-gneiss, and migmatite gneiss samples obtained from a depth below 1251 m were subjected to laboratory measurements. A variable compressive strength of 22–98 MPa was exhibited by the granitoids samples. These included granite (41–98 MPa), granite-gneiss (22–88 MPa), and migmatite gneiss (41–67 MPa). In an earlier study, Singh (1981) provided a compilation of uniaxial compressive strength of 690 granitic rocks with varying composition for several sites globally. Figure 7.37 shows a comparison of unconfined compressive strength of Koyna granitoids

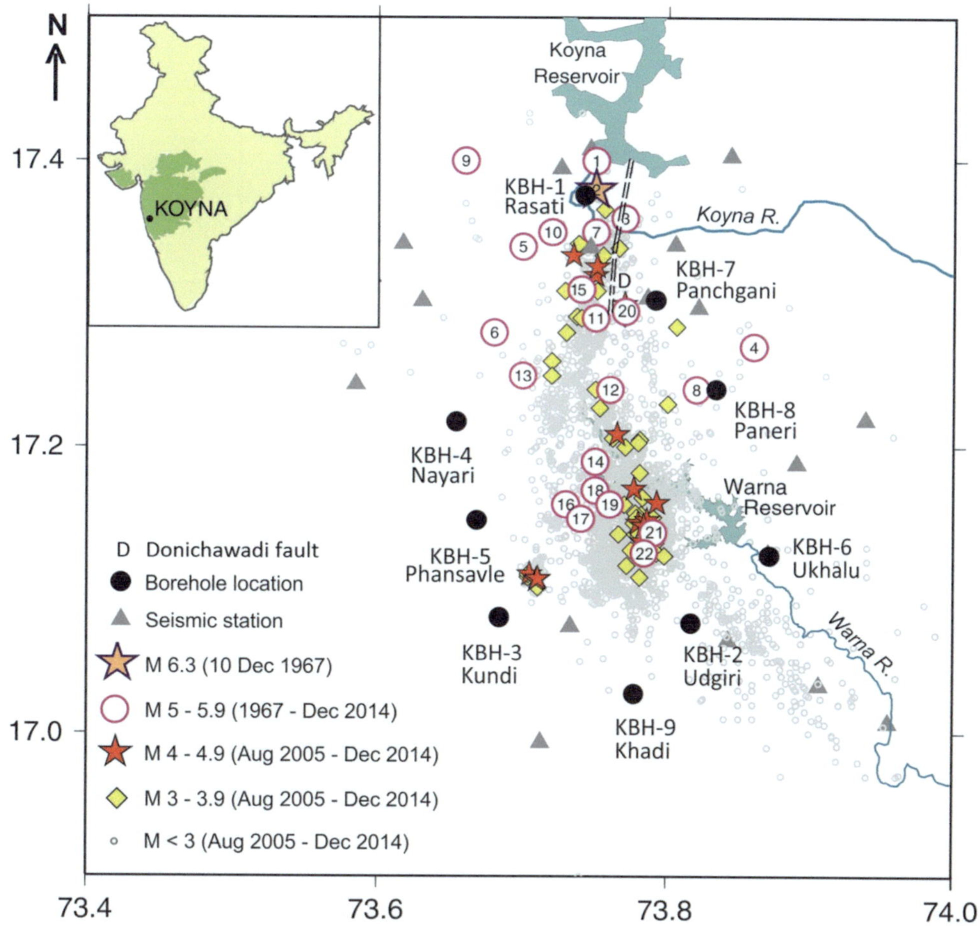

Fig. 7.35 Map of Koyna-Warna area showing the distribution of cored boreholes vis-a-vis the seismicity (modified after Gupta et al. 2015; Goswami et al. 2017a). The Koyna and Warna artificial water reservoirs are also shown. The trace of the Donachiwada fault zone, formed during 1967 M 6.3 Koyna earthquake, is also shown. Inset shows the location of Koyna in the Deccan Volcanic Province (shaded green) on the outline map of India

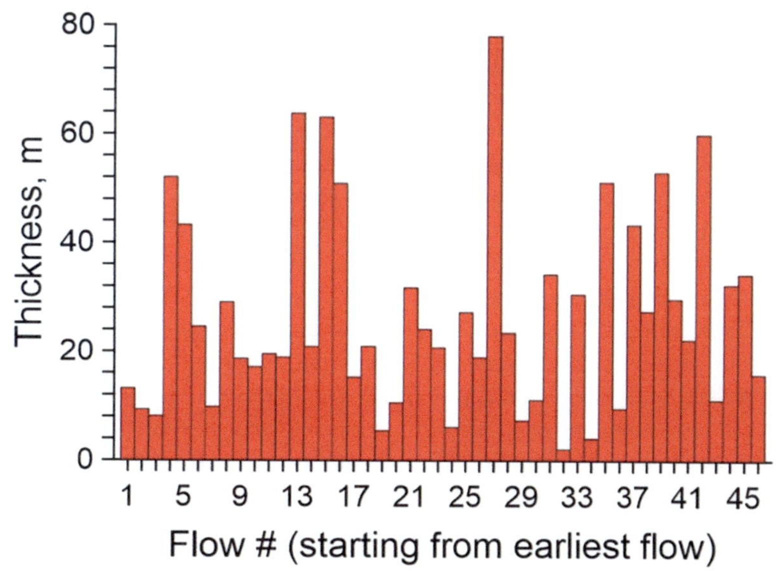

Fig. 7.36 Bar graph showing the thicknesses of individual lava flows identified in borehole KBH-7. Flows are numbered with the first flow directly overlying the granitic basement designated as Flow # 1

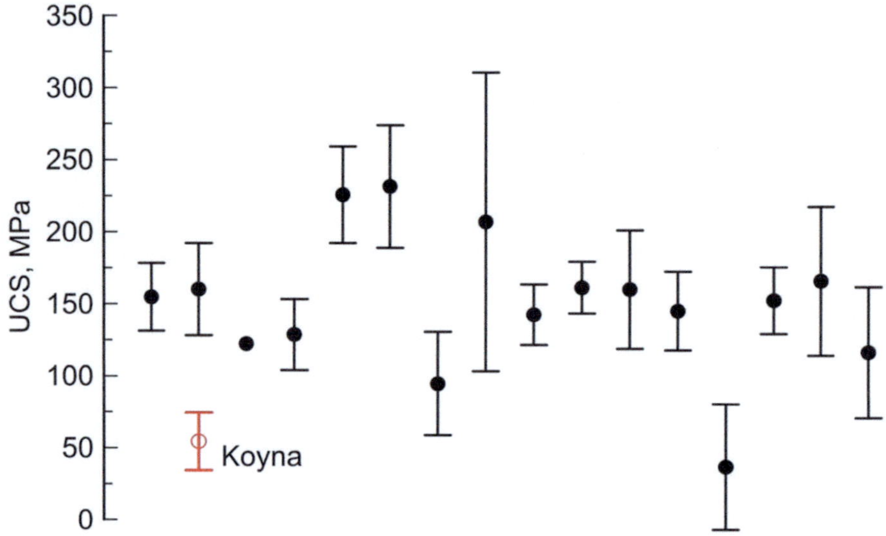

Fig. 7.37 Comparison of mean UCS of granitoids from Koyna area (red open circle) and mean UCS of granitic rocks from 16 sites (690 samples) reported in the compilation by Singh (1981). The bars indicate standard deviation

with granitic rock samples from 16 sites globally from Singh (1981). In general, Koyna samples have much lower unconfined compressive strength. In another study, Goswami et al. (2017b) have examined samples from KBH-3, KBH-4, KBH-5 and KBH-7 under varying confining pressure of 6 MPa, and obtained relatively low compressive strength of samples from the Koyna region. From these studies it may be concluded that the comparatively low compressive strength and elastic properties of the basement rocks compared to other regions may be due to the recurrent seismic activity. Low rock strength implies that the media is not capable of holding strains to generate significantly strong earthquakes. The accumulated strains get released through smaller earthquakes.

7.27 A Few Recent Studies

Several important studies have been conducted in the recent years elucidating the rock properties, status of stress field and related topics in the Koyna region. These are too many to be covered in this book. In the following, a few of these are included. Dutta et al. (2018) undertook exploration of deep terrestrial subsurface microbiome in the Deccan traps of Late Cretaceous age and the underlain Archean basement. They observed that the deep borehole provided a unique opportunity to explore the microbial life in the deep biosphere hosted by the ~ 65-million-year-old Deccan Basalt. The abundance of autotrophic and heterotrophic taxa is strongly correlated. They inferred that possible community isolation could be due to the lack of shared operational taxonomic units and distinct clustering of major taxa. They concluded that geomicrobiological studies carried out in Koyna deep borehole provide excellent opportunities to comprehend microbial composition and function in the terrestrial igneous rock-hosted deep biosphere. Goswami et al. (2019) delineated the damaged zones in the 3-km-deep Pilot Borehole. The borehole went through 1247 m of the Deccan Basalt cover and penetrated 1767 m into the granitic basement. Physical properties and acoustic behaviour of the granitic basement were investigated using geophysical logs from 1500 to 3000 m depth. Major findings included identification of 7 anomalous zones below 2100 m depth on the bases of electrical resistivity, caliper, density, neutron porosity, self-potential and sonic logs. The anomalous zones are characterized by shear wave velocity anisotropy, which is found to be primarily stress induced. Comparison of fast-polarized shear wave azimuth (FSA) with independent estimates of SH max orientation inferred from drilling-induced tensile fractures located in the anisotropic zone led to the conclusion that FSA is mainly controlled by the stress regime. In yet another interesting study, Goswami et al. (2020) have presented in situ stress orientation in the 3-km-deep Pilot Borehole and commented on the transitional faulting environment. The orientation of in situ horizontal stress is determined from the stress-induced wellbore failures, breakouts, and tensile fractures induced by drilling identified in the acoustic and micro-resistivity images of the granitic basement.

Major inferences drawn by Goswami et al. (2020) include N9° W $\pm$ 17° orientation for the SH max horizontal principal stress, which is consistent with the regional NNW–SSE orientation of SH max. It is inferred from the breakout rotations at several depths below 2100 m, the borehole must have passed through localized fault damage zones. The conditions in the borehole indicate favorable stress orientation to support strike–slip and normal faulting. Podugu et al. (2019) have reported noble gas isotope composition of the formation gases from the Pilot Borehole. From depths of 1315–2831 m samples were collected every ~ 100 m to study CO_2, CH_4, H_2, and He. The concentrations measured were 1200 parts per million by volume (ppmv) for CO_2, 186 ppmv for CH_4,

139 ppmv for H_2, and 12.8 ppmv for He. Zones below 2100 depth are normally rich in these gas concentrations and correlate well with the observed fractured zones in borehole. One significant result is that from the study of Helium isotope ratio, it is inferred that helium found in the borehole is composed of atmospheric and crustal radiogenic component and not derived from the mantle.

Work is under different stages of progress with the aim of placing the ~ 7-km-deep bore-hole laboratory for the near-field studies of RTS events at Koyna.

References

Ando M (2001) Geological and geophysical studies of the Nojima Fault from drilling: an outline of the Nojima Fault Zone Probe. Island Arc 10:206–214

Anuradha M, Mallika K (2017) Reservoir loading parameters at Koyna and their correspondence with reservoir triggered seismicity. J Geol Soc India 90(6):802–804

Arora K, Chadha RK, Srinu Y, Selles A, Davuluri S, Smirnov V, Ponomarev A, Mikhailov VO (2017) Lineament fabric from airborne LiDAR and its influence on triggered earthquakes in the Koyna-Warna region, western India. J Geol Soc India 90(6):670–677

Auden JB (1952) Geological inspection note, Koyna hydroelectric project, Bombay state: Geological Survey of India, unpublished report

Bansal AR, Rao NP, Peng Z, Shashidhar D, Meng X (2018) Remote triggering in the Koyna-Warna reservoir-induced seismic zone western India. J Geophys Res (Solid Earth) 123(3):2318–2331

Beane JE, Turner CA, Hooper PR, Subbarao KV, Walsh JN (1986) Stratigraphy, composition and form of the Deccan Basalts, Western Ghats, India. Bull Volcanol 48:61–83

Bodas MS, Khadri SFR, Subbarao KV (1988) Stratigraphy of the Jawhar and Igatpuri formations, western Deccan Basalt Province. In: Subbarao KV (ed) Deccan flood basalts. Geological Society of India, Memoirs, vol 10, pp 235–252

Bondre NR, Duraiswami RA, Dole G (2004) Morphology and emplacement of flows from the Deccan Volcanic Province, India. Bull Volcanol 66:29–45

Chandra U (1976) Focal mechanism of the Koyna, India earthquake of 1967, December 10. Geophys J Int 46:247–251

Cox KG, Hawkesworth CJ (1985) Geochemical stratigraphy of the Deccan Traps at Mahabaleshwar, Western Ghats, India, with implications for open system magnetic processes. J Petrol 26:355–377

Das D, Mallik J (2020) Koyna earthquakes: a review of the mechanisms of reservoir-triggered seismicity and slip tendency analysis of subsurface faults. Acta Geophys 68:1097–1112. https://doi.org/10.1007/s11600-020-00457-6

Devey CW, Lightfoot PC (1986) Volcanological and tectonic control of stratigraphy and structure in the Western Deccan Traps. Bull Volcanol 48:195–207

Dodge DA, Beroza GC, Ellsworth WL (1996) Detailed observations of California foreshock sequences: implications for the earthquake initiation process. J Geophys Res 101:22371–22392

Dutta A, Gupta SD, Gupta A, Sarkar J, Roy S, Mukherjee A, Sar P (2018) Exploration of deep terrestrial subsurface microbiome in Late Cretaceous Deccan traps and underlying Archean basement, India. Sci Rep 8(1):1–16

Elliot JR, Michele M de, Gupta HK (2020) Earth Observation for Crustal Tectonics and Earthquake Hazards. Surveys in Geophysics 41:1355–1389.

Goswami D, Akkiraju VV, Singh SK, Shalivahan SR (2017a) Rock strength and elastic properties of basement granitoids from Koyna region, Deccan Volcanic Province, India. J Geol Soc India 90(6):783–787

Goswami D, Akkiraju VV, Misra S, Roy S, Singh SK, Sinha A, Gupta HK, Bansal BK, Nayak S (2017b) Rock strength measurements on Archaean basement granitoids recovered from scientific drilling in the active Koyna seismogenic zone, western India. Tectonophysics 182–192

Goswami D, Roy S, Akkiraju VV (2019) Delineation of damage zones from 3 km downhole geophysical logs in the Koyna seismogenic zone, western India. J Geophys Res Solid Earth 124(6):6101–6120

Goswami D, Hazarika P, Roy S (2020) In situ stress orientation from 3 km borehole image logs in the Koyna seismogenic zone western India: implications for transitional faulting environment. Tectonics 39(1):e2019TC00564747

GSI (1968) A geological report on the Koyna earthquake of 11th December, 1967, Satara district, Maharashtra state. Report submitted by officers of the Geological Survey of India, Mar 1968, 273 p (incl. plates)

Gupta HK (1992) Reservoir induced earthquakes. Elsevier Scientific Publishing Company, Amsterdam, p 364

Gupta HK (2001) Short-term earthquake forecasting maybe feasible at Koyna, India. Tectonophysics 338(3–4):353–357

Gupta HK (2002) A review of recent studies of triggered earthquakes by artificial water reservoirs with special emphasis on earthquakes in Koyna, India. Earth-Sci Rev 58(3–4):279–310

Gupta HK (2017) Koyna, India, an ideal site for near field earthquake observations. Geol Soc India 90(6):645–652

Gupta HK (2018) Review: reservoir triggered seismicity (RTS) at Koyna, India over the past 50 years. Bull Seismol Soc Am 108(5B):2907–2918

Gupta HK (2021) Artificial water reservoir triggered earthquakes with special emphasis on Koyna, India. In: Encyclopedia of solid earth geophysics, vol 2, 2nd edn. Springer + Business Media, pp 19–32

Gupta HK (2022) Artificial water reservoir-triggered seismicity (RTS): most prominent anthropogenic seismicity. Surv Geophys. https://doi.org/10.1007/s10712-021-09675-z

Gupta ML, Gaur VK (1984) Surface heat flow and probable evolution of Deccan volcanism. Tectonophysics 105:309–318

Gupta HK, Nayak S (2011) Deep scientific drilling to study reservoir-triggered earthquakes in Koyna, western India. Sci Drill 12:53–54

Gupta HK, Rastogi BK (1974) Will another damaging earthquake occur in Koyna? Nature 248:215–216

Gupta HK, Rastogi BK, Narain H (1972) Common features of the reservoir associated seismic activities. Bull Seismol Soc Am 62:481–492

Gupta HK, Rao CVR, Rastogi BK, Bhatia SC (1980) An investigation of earthquakes in Koyna region, Maharashtra, for the period October 1973 through December 1976. Bull Seismol Soc Am 70:1833–1847

Gupta HK, Rao RUM, Srinivasan R, Rao GV, Reddy GK, Dwivedi KK, Banerjee DC, Mohanty R, Satyasaradhi YR (1999) Anatomy of surface rupture zones of two stable continental region earthquakes, 1967 Koyna and 1993, Latur, India. Geophys Res Lett 26(13):1985–1988

Gupta HK, Mandal P, Rastogi BK (2002) How long will triggered earthquakes at Koyna, India continue? Curr Sci 82(2):202–210

Gupta HK, Shashidhar D, Periera M, Mandal P, Rao NP, Kousalya M, Satyanarayana HVS, Dimri VP (2007) Earthquake forecast appears feasible at Koyna, India. Curr Sci 93:843–848

Gupta HK, Shashidhar D, Mallika K, Purnachandra Rao N, Srinagesh D, Satyanarayana HVS, Saha S, Naik RTB (2011) Short term earthquake forecasts at Koyna, India. J Geol Soc India 77:5–11

Gupta HK, Nayak S, Ellsworth W, Rao YJB, Rajan S, Bansal BK, Purnachandra Rao N, Roy S, Arora K, Mohan R, Tiwari VM, Satyanarayana HVS, Patro PK, Shashidhar D, Mallika K (2014) Probing reservoir-triggered earthquakes in Koyna, India, through scientific deep drilling. Sci Drill 18(5–9):2014

Gupta HK, Purnachandra Rao N, Roy S, Arora K, Tiwari VM, Patro BPK, Satyanarayana HVS, Shashidhar D, Mallika K, Akkiraju VV, Goswami D, Vyas D, Ravi G, Srinivas KNSSS, Srihari M, Mishra S, Dubey CP, Raju CV, Borah U, Chinna Reddy K, Babu N, Rohilla S, Dhar U, Sen M, Bhaskar Rao YJ (2015) Investigations related to scientific deep drilling to study reservoir-triggered earthquakes at Koyna, India. Int J Earth Sci 104:1511–1522

Gupta HK, Arora K, Purnachandra Rao N, Roy S, Tiwari VM, Patro PK, Satyanarayana HVS, Shashidhar D, Mahato CR, Srinivas KNSSS, Srihari M, Satyavani N, Srinu Y, Gopinadh D, Raza H, Jana M, Akkiraju VV, Goswami D, Vyas D, Dubey CP, Raju DCV, Borah U, Raju K, Chinna Reddy K, Babu N, Bansal BK, Nayak S (2016) Investigations of continued reservoir triggered seismicity at Koyna, India. Geol Soc Lond Spec Publ 445:151–188

Hazarika P, Yadav A, Roy S (2017) Influence of permeability in modeling of reservoir triggered seismicity in Koyna region, western India. J Geol Soc India 90(6):728–732

Helmestetter A, Sornette D, Grasso JR (2003) Mainshocks are aftershocks of conditional foreshocks: how do foreshock statistical properties emerge from aftershock laws? J Geophys Res 108B:2046

Hickman S, Zoback MD, Younker L, Ellsworth W (1994) Deep scientific drilling in the San Andreas fault zone. Eos Trans AGU 75:137, 140, 142

Johnston AC, Kanter LR, Coppersmith KJ, Cornell CA (1994) The earthquakes of stable continental regions. In: Volume 3: seismicity data sheets (part 1); final report no. EPRI-TR-102261-V3. Electric Power Research Institute

Kaiser J (1953) Arch Eisenhuettenwes 24:43–45 [in German]

Khadri SFR, Subbarao KV, Hooper PR, Walsh JN (1988) Stratigraphy of Thakurvadi Formation, Western Deccan Basalt Province, India. In: Subbarao KV (ed) Deccan flood basalts. Geological Society of India, Memoirs, no 10, pp 281–304

Krishnan S, Crosby CJ, Nandigram V, Phan M, Cowart C, Baru C (2011) Open topography: a service-oriented architecture for community access to LiDAR topography. https://doi.org/10.1145/1999320.1999327

Langston CA (1976) A body wave inversion of the Koyna, India earthquake of December 10, 1967, and some implications for body wave focal mechanisms. J Geophys Res 81:2517–2529

Langston CA (1981) Source inversion of seismic waveforms: the Koyna, India, earthquakes of 13 September 1967. Bull Seismol Soc Am 71:1–24

Ma K-F, Tanaka H et al (2006) Slip zone and energetics of a large earthquake from the Taiwan Chelungpu fault drilling project. Nature 444:473–476. https://doi.org/10.1038/nature05253

Madugula UA, Rekapalli R, Sudheer K, Jagadeesh P, Satyanarayana HVS (2021) Triggering of earthquakes at Koyna, India and hammer drilling for the pilot borehole. Nat Hazards 109:1191–1199

Mallika K, Gupta H, Shashidhar D, Purnachandra Rao N, Yadav A, Rohilla S, Satyanarayana HVS, Srinagesh D (2013) Temporal variation of b value associated with M ~ 4 earthquakes in the reservoir-triggered seismic environment of the Koyna–Warna region, western India. J Seismol 17:189–195

Mishra S, Misra S, Vyas D, Nikalje D, Warhade A, Roy S (2017) A 1251 m-thick Deccan flood basalt pile recovered by scientific drilling in the Koyna region, western India. J Geol Soc India 90(6):788–794

Misra S, Bartakke V, Athavale G, Akkiraju VV, Goswami D, Roy S (2017) Granite-gneiss basement below Deccan Traps in the Koyna region, western India: outcomes from scientific drilling. J Geol Soc India 90(6):776–782

Mitchell C, Cox KG (1988) A geological sketch map of the southern part of the Deccan Province. In: Subbarao KV (ed) Deccan flood basalts. Geological Society of India, Memoirs, vol 10, 393 pp

Ohnaka M (1992) Earthquake source nucleation: a physical model for short-term precursors. Tectonophysics 211:249–278

Podugu N, Mishra S, Wiersberg T, Roy S (2019) Chemical and noble gas isotope compositions of formation gases from a 3 km deep scientific borehole in the Koyna seismogenic zone, western India. Geofluids 2019:1–16

Raman R, Majumdar S, Dubey R (1984) Sequence of basaltic flows around Koyna dam, Satara district, Maharashtra. In: Symposium on Deccan Trap and bauxite (1976). Geological Survey of India Special Publication No. 14, pp 15–17

Rao NP, Shashidhar D (2017) Earthquake focal mechanism studies in Koyna-Warna region in the last five decades—current understanding on tectonics and seismogenesis. J Geol Soc India 90(6):684–691

Rao NP, Arora K, Ponomarev A, Nascimento AF (2018) Fifty years after the 1967 Koyna earthquake—lessons learned about reservoir triggered seismicity (RTS). Introduction to the special issue. Bull Seismol Soc Am 108(5B):2903–2906

Roy S, Rao NP et al (2013a) Granitic basement below Deccan Traps unearthed by drilling in the Koyna Seismic Zone, western India. J Geol Soc India 81:289–290

Sangode SJ, Venkateshwarulu M, Mahajan R, Randive V (2017) Magnetic mineralogical variability along Deccan Trap basalt borehole (KBH07), Koyna deep continental drilling program, western Maharashtra, India. J Geol Soc India 90(6):769–775

Shashidhar D, Mallika K, Rao NP, Gupta HK, Satyanarayana HVS (2014) Detection of quarry blasts in the Koyna-Warna region, western India. OJER 2014(3):162–169

Shashidhar D, Satyanarayana HVS, Mahato CR, Mallika K, Purnachandra Rao N, Gupta HK (2016) Borehole seismic network at Koyna, India. Seismol Res Lett 87(3):661–667

Shashidhar D, Mallika K, Mahato CR, Maity BS, Satyanarayana HVS (2017) Recent seismicity patterns and microearthquake activity on an active intraplate fault system at Koyna-Warna, western India. J Geol Soc India 90(6):798–801

Simpson DW, Narasimhan TN (1990) Inhomogeneities in rock properties and their influence on reservoir induced seismicity. Gerlands Beitr Geophys 99:205–219

Singh MM (1981) Strength of rock. In: Toulakian YS, Judd WR, Roy RF (eds) Physical properties of rocks and minerals. McGraw-Hill/CINDAS data series on material properties, vol II-2, pp 83–121

Singh DD, Rastogi BK, Gupta HK (1975) Surface-wave radiation pattern and source parameters of the Koyna earthquake of December 10, 1967. Bull Seismol Soc Am 65:711–731

Srinivasan PB, Dutta AK (1960) Progress report on Koyna hydroelectric project, North Satara district, Maharashtra: 1959–60. Geological Survey of India, unpublished report

Srinivasan PB, Dutta AK (1961) Progress report for 1960–61 on Koyna hydroelectric project, North Satara district, Maharashtra. Geological Survey of India, unpublished report

Subbarao KV, Courtillot V (2017) Deccan Basalts in and around Koyna-Warna region, Maharashtra—some reflections. J Geol Soc India 90(6):653–662

Subbarao KV, Hooper PR (1988) Reconnaissance map of the Deccan Basalt Group in the Western Ghats, India. In: Subbarao KV (ed) Deccan flood basalts. Geological Society of India, Memoirs, vol 10, 393 pp

Talwani P, Kumaraswamy SV, Sawalwade CB (1996) Re-analysis of earthquakes of Koyna-Warna, 1963–1995. MERI report

Walker GPL (1971) Simple and compound lava flows and flood basalts. Bull Volcanol 35:1–12

Widdowson M, Mitchell C (1999) Large-scale stratigraphical, structural and geomorphological constraints for earthquakes and seismicity in the southern Deccan Traps, India. The case for denudationally driven seismicity. In: Subbarao KV (ed) Deccan Volcanic Province. Geological Society of India, Memoirs, vol 43, no 1, pp 425–452

Yadav A, Shashidhar D, Mallika K, Rao NP, Rohilla S, Satyanarayana HVS, Srinagesh D, Gupta H (2013) Source parameters of earthquakes in the reservoir-triggered seismic (RTS) zone of Koyna-Warna, western India. Nat Hazards 69:965–979

Yadav A, Bansal BK, Pandey AP (2017) Five decades of triggered earthquakes in Koyna-Warna region, western India—a review. Earth Sci Rev 162:433–450

Zeilke O, Arrowsmith JR, Ludwig LG, Akciz SO (2012) High-resolution topography-derived offsets along the 1857 Fort Tejon earthquake rupture trace, San Andreas fault. Bull Seism Soc Am 102(3):1135–1154

Zoback MD, Hickman S, Ellsworth WL (2010) Scientific drilling into the San Andreas fault zone. Eos 91:197–198

Zoback MD, Hickman S, Ellsworth WL (2011) Scientific drilling into the San Andreas fault zone—an overview of SAFOD's first five years. Sci Drill 11:14–28. https://doi.org/10.2204/iodp.sd.11.02.2011

8 Injection Induced/Triggered Seismicity in Oklahoma, USA

In a review entitled 'Injection Induced Earthquakes', Ellsworth (2013) has addressed the occurrence of earthquakes in unusual locations in America and Europe. This has raised concerns about the anthropogenic activities that have been responsible for triggering such events. He has particularly focused attention on injection induced earthquakes. He has highlighted the occurrence of the 6 November 2011 Mw 5.7 Oklahoma, USA earthquake considered to have been triggered by fluid injection for enhanced oil recovery. Among the measures to reduce the risk of injection induced earthquakes, he has advocated the application of the suggestion made by Zoback (2012), which involves setting up of '*traffic light*' system. Ellsworth (2013) has further pointed out the failure of this approach as witnessed in Basel, Switzerland, which failed in preventing Mw 3 earthquakes, but may have prevented larger earthquakes (Deichmann and Giardini 2009).

It is important for the completeness of the studies on injection triggered seismicity to address briefly the very first studies conducted at Denver, Colorado, USA and Rangely Oil Field also located in Colorado, USA. A deep disposal well was drilled to a depth of 3671 m at the Rocky Mountain Arsenal located northeast of Denver Colarodo (Evans 1966). The disposal of the waste fluid started on the 8 March 1962 that continued till 30 September 1963. The rate of amount of fluid disposed was 21 million L monthly. After a gap of 10 months fluid injection under gravity flow started in August 1964, on a monthly average of 7.5 million L and continued till 6 April 1965. This gravity flow disposal was followed by under pressure disposal of 17 million L per month. Within a few weeks of the initiation of fluid disposal on 8 March 1962, a swarm of tremors started in April 1962, in the vicinity of the disposal well. Consequently, disposal of the waste fluids was terminated on 20 February 1966. Colorado is known to be an area of low seismicity. 30 felt earthquakes have been listed by Hadsell (1968) in the Colorado region in past 100 years before the initiation of disposal. Prior to initiation of fluid injection in 1962, only one earthquake in the area is known to have occurred in 1882 (Simon 1969). A distinct correlation between injection of fluids and occurrence of earthquakes has been established by Evans (1966). Figure 8.1, after Handin and Raleigh (1972) gives the monthly number of earthquakes in the Denver area, monthly volume of the fluid injected and wellhead pressure. Figure 8.2 provides the epicenters located in the immediate vicinity of the Rocky Mountain Arsen Well in January and February 1966 using a dense network of seismic stations (after Healy et al. 1968). After stopping of the injection in early 1966, no tremors were reported. A year later three earthquakes in the magnitude range of M 5.0–5.2 occurred. The total number of earthquakes is estimated to be 1500 during the period 1962–1967. The studies at the Rocky Mountain Arsenal Well triggered earthquakes provided the initial insight to the phenomenon of induced seismicity (Healy et al. 1968).

The Rangely Oil Field in Colorado, USA is another early example of earthquakes being induced due to water injection to enhance oil production. In the Uinta Basin in Colorado, oil had been produced from a closed anticline. The production started in 1945. In the years to follow, the formation of fluid pressure decreased with the production of oil. To enhance the oil production water was injected in 1957. Later in 1962, a seismological observatory was installed in a location, which is about 65 km west-northwest of Rangely. Tremors were immediately reported from the close vicinity of the Rangely Oil Field (Raleigh 1972). A correlation in number of earthquakes and the volume of fluid injected was observed (Gibbs et al. 1973). This necessitated establishment of a local network of portable seismographs (Pakiser et al. 1969). Earthquakes were found to be clustered in two groups (Fig. 8.3), where pore pressure due to water injection exceeded the local fluid pressure. Deployment of a local network of 14 permanent seismographs during 1969/

H. K. Gupta, *Artificial Water Reservoir Triggered Seismicity*, https://doi.org/10.1007/978-3-032-05873-7_8

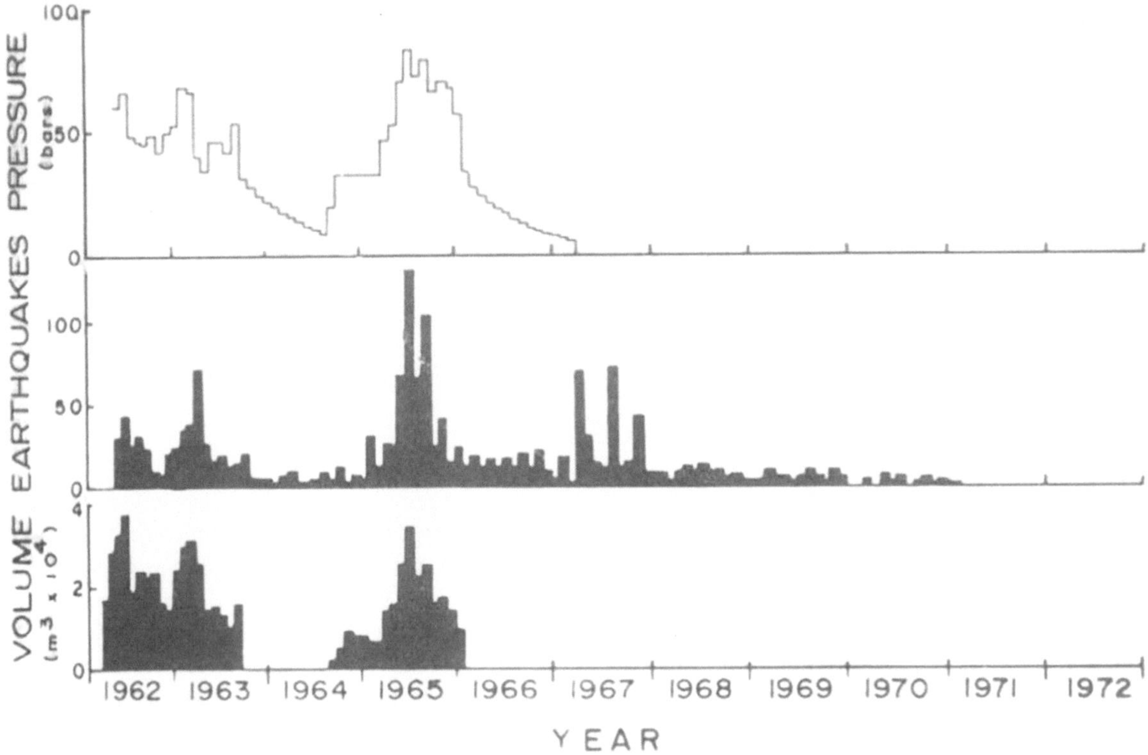

Fig. 8.1 Monthly number of earthquakes in the Denver area, monthly volume of injected water and wellhead pressure in the disposal well (after Handin and Raleigh 1972)

Fig. 8.2 Epicenters of earthquakes near Denver deep-disposal well located in January and February 1966, by means of a dense network of seismic stations (after Healy et al. 1968)

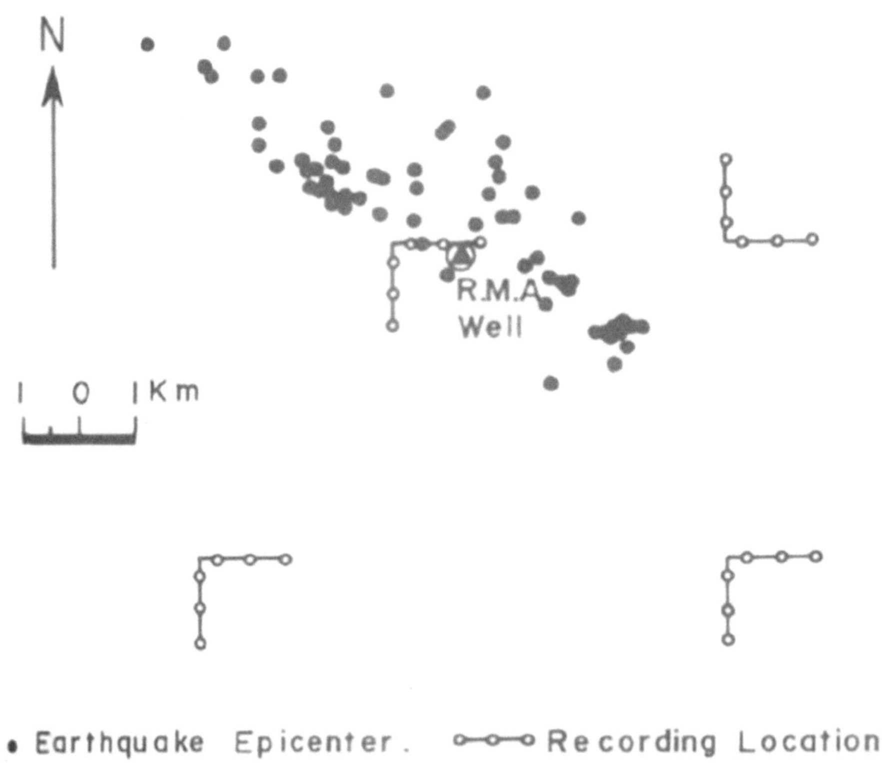

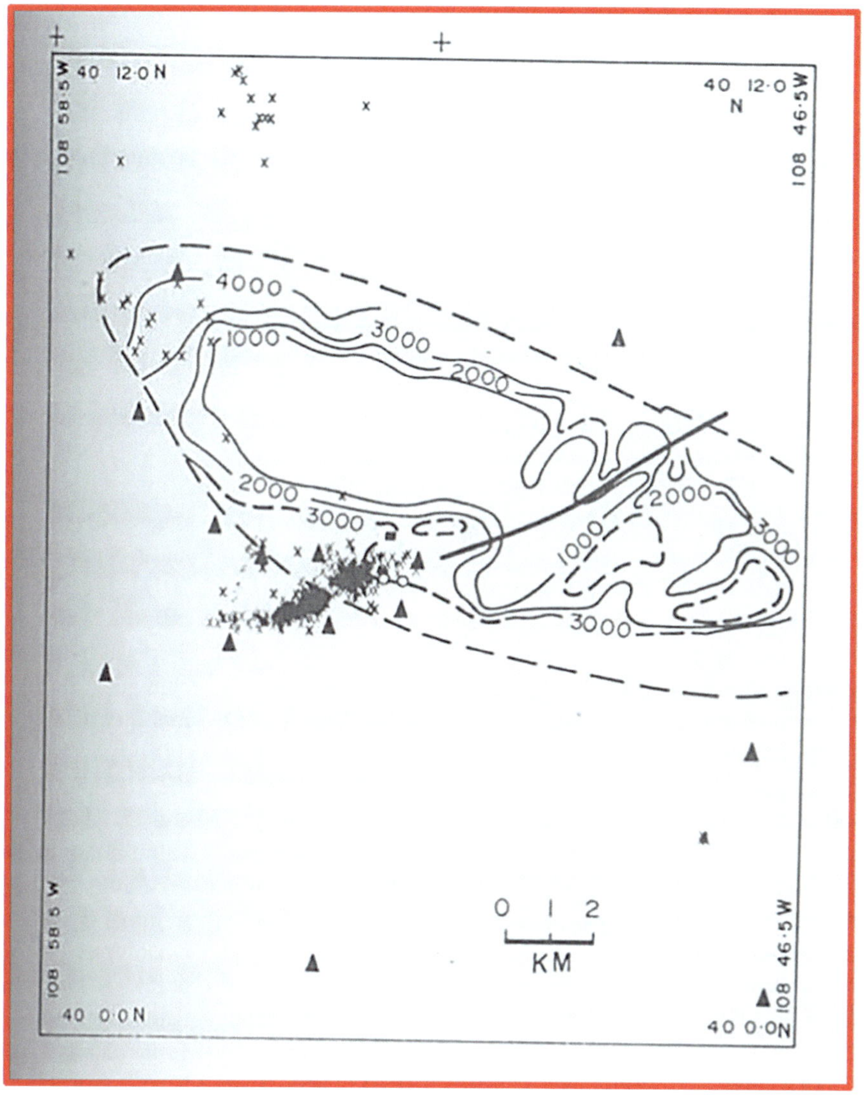

Fig. 8.3 Epicenters of earthquakes in the Rangely oil field (outlined by the dashed line) between October 1969, and November 1970. Contours give the bottom–hole pressure (in psi) measured in September 1969. Triangles represent seismograph stations; circles, experimental wells; square, well in which stresses were measured by means of hydraulic fracturing (after Raleigh 1972)

1970 improved the location of the earthquakes. The earthquakes were located mostly from the south-central part of the Rangely Oil Field at the top of the Weber sandstone where the pore pressure created by the injection was greater than the original field pressure.

The Rangely Oil Field in Colorado provided another early example of earthquake being induced by fluid injection.

In this chapter, the characteristics of injection induced earthquakes in the north-central Oklahoma are discussed and it is found that these characteristics are like reservoir triggered seismicity (RTS) as inferred by Rajesh and Gupta (2021).

8.1 Earthquakes in Oklahoma, USA

It has been well established that prior to 2009, Oklahoma was characterized by a low level of seismicity and was considered to be a stable continental region, as inferred from instrumental earthquake records as well as low intra-plate strain rates estimated from GPS observations as mentioned by Calais et al. (2006) and Hough and Page (2015).

Earthquakes of M 5.0, 5.7 and 5.0 occurred near Prague, Oklahoma on 5, 6 and 8 November 2011 respectively. This was indeed unusual for the region that was known to have slow strain rate accumulation (Keranen et al. 2013, 2014). Later Pawnee earthquake of M 5.8 occurred on 3 September 2016. It may be noted that the increase in seismicity in Oklahoma was noticed in 2009, some 20 years after the initiation of injection of wastewater into saltwater disposal (SWD) wells, to increase oil production, and 5 years after the substantial increase in wellhead pressure (Horton 2012; Schoenball and Ellsworth 2017; Van der Baan and Calixto 2017). It may be noted that the volume of the wastewater disposed increased almost two-fold between 2004 and 2008 which coincided with 900-fold increase in the annual number of M $\geq$ 3 earthquakes by 2015 (Keranen et al. 2014). Subsequently, the injection rates were reduced in 2016 and that was followed by a remarkable decrease in seismicity (Langenbruch and Zoback 2016; Johann et al. 2018).

The maximum estimated magnitude of earthquake is M 6 for the Oklahoma region, as estimated from the known seismic activity for the region for the past several centuries (Van der Elst et al. 2016). The Pawnee earthquake of 3 September 2016 had a magnitude of M 5.8. This resulted in reducing the volume of fluids being injected by 40% of the volume being injected during January 2014 and March 2016 by the Oklahoma Regulators. A decrease in the number of earthquakes was subsequently observed. As compared to ~ 900 M $\geq$ 3 earthquakes during 2015, there were 624 and 304 such events in 2016 and 2017. This number further dropped to 194 in 2018 (Walter et al. 2019). Modeling of injection induced seismicity (IIS) and reservoir triggered seismicity carried out by Johann et al. (2018) have shown that the level of seismicity depends upon the prevalent level of strain accumulation of the host rocks and the volume and the depth of injection. This shall be helpful in planning future fluid disposal sites.

Gupta et al. (1972a, b) have examined in detail the 'b' values and foreshock/aftershock patterns, ratio of the largest aftershock to the main shock magnitude and aftershocks decay patterns to classify earthquake sequences in the three types of earthquake sequences discussed by Mogi (1962, 1967). This has been presented in Chap. 3 of the present book. It was considered useful to examine the Oklahoma seismicity considering Mogi's classification (Rajesh and Gupta 2021).

8.2 Analysis of Earthquakes in Oklahoma, USA

Rajesh and Gupta (2021) made use of the earthquake catalogue of Oklahoma Geological Survey (OGS) for the period 1 January 2002 through 31 December 2018. OGS catalogue uses M_L and Mw (USGS). These were converted to M using a regression relationship (Rajesh and Gupta 2021). Figure 8.4 depicts the spatio-temporal distribution of M $\geq$ 3 earthquakes for the period 2002 through 2018. The catalogue was found to be complete for M 2.1. In the Gutenberg Richter relation:

$$\log_{10} N = A - b\,M$$

where N is the number of events with magnitude $\geq$ M, and A and b-values are 7.18 and 1.31, for the period 2002 through 2018 (Fig. 8.5). Rajesh and Gupta computed b values for different periods as indicated in Fig. 8.5. The observed increase in b values post 2009 could be due to pore pressure changes caused by injections from different wells (Bachmann et al. 2012). It may be noted that the b value of 1.00 during the pre-injection period of 2002–2009 is lower than the 1.28 and 1.48 observed later for the periods 2010–2016 and 2017–2018 respectively. The temporal variation of b-value computed using a window length of 6 h along with seismicity is depicted in Fig. 8.6. A drop in the increasing trend of b values preceding the occurrence of main shocks is obvious in Fig. 8.6. Similar trends of drop in b values before the occurrence of M 3 and larger earthquakes has been reported for Koyna, India (Mallika et al. 2013). The b value changes associated with the Prague M 5.7 and Pawnee M 5.8 earthquakes are zoomed from Fig. 8.6 in Fig. 8.7, where a sharp fall in b value before the two earthquakes is very prominent. A decrease in b value could be used as signal for a forthcoming earthquake (Main et al. 1989).

Table 8.1 provides the b values for foreshocks and aftershocks, magnitude ratio of the largest aftershock to the main shock, difference of the magnitude between the mainshock and its largest aftershock for the Prague M 5.7 and Pawnee M 5.8 earthquakes. The aftershocks are fitted using Utsu (1961) relation:

$$n(t) = Ct^{-h}$$

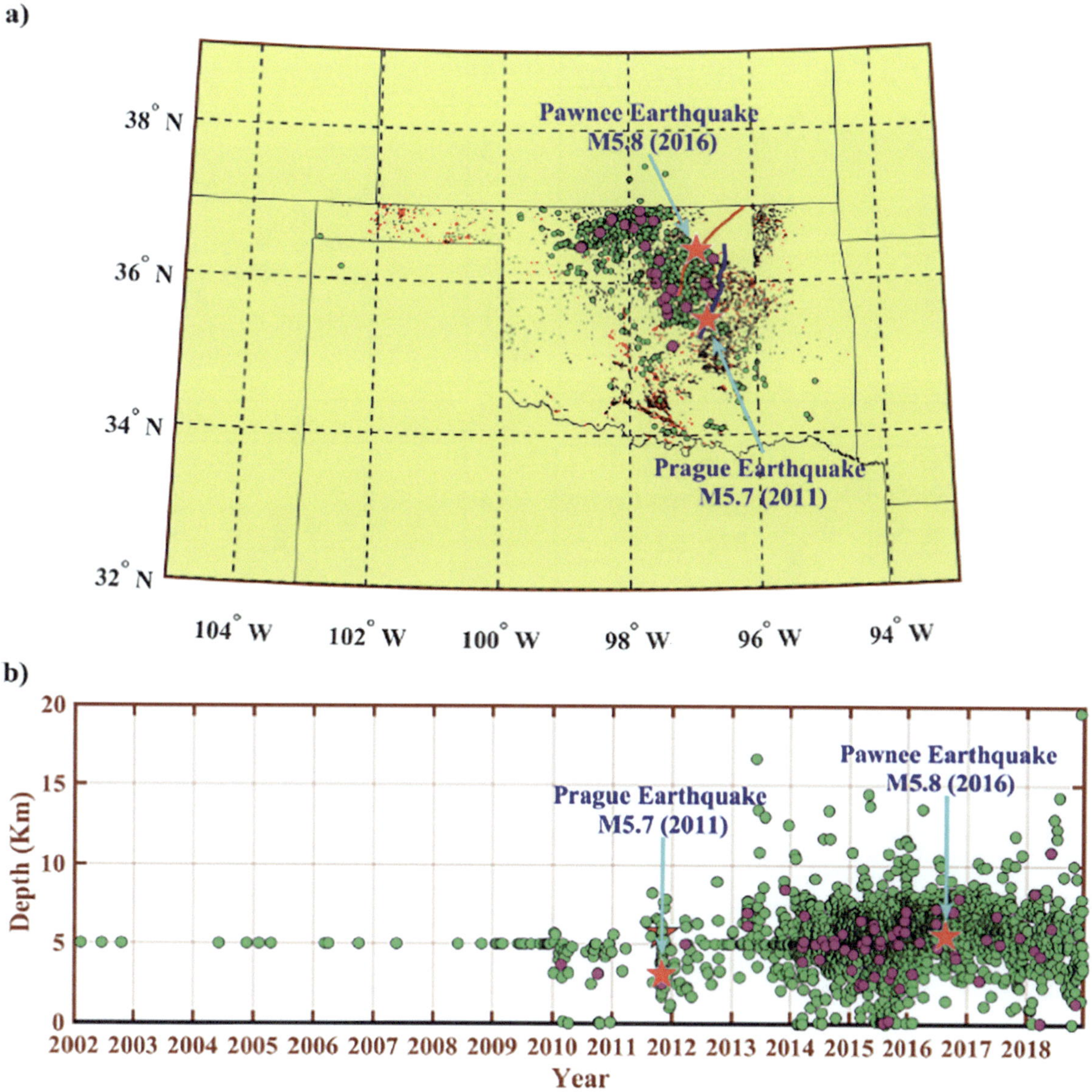

Fig. 8.4 Distribution of earthquake with magnitude $3 \leq M < 4$ earthquakes (green circles), $4 \leq M \leq 5.5$ (pink circles) at Oklahoma from January 2002 to December 2018. Red stars denote the Prague (M 5.7) and Pawnee (M 5.8) earthquakes whose sequences are analyzed. **a** Wastewater disposal wells and EOR and production wells are shown with red and black color dots, respectively, and spatial distribution of earthquakes of magnitude $\geq$ 3. The Wilzetta and Labette faults are shown with blue and brown color lines. **b** Depth distribution of events during 2002–2018

where n(t) is the aftershock frequency per unit time, C and *h* are constants. Using the approach outlined by Gupta (1992), $M_0 - M_1$, and M_1/M_0 are computed. M_0 and M_1 are the magnitudes of the main shock and the largest aftershock. The estimates are provided in Table 8.1 for C, h, and $M_0 - M_1$.

The foreshock aftershock patterns for Prague and Pawnee earthquakes are shown in Fig. 8.8. Both these earthquakes are shown to be followed by aftershocks. The foreshock aftershock pattern belongs to Type 2 model.

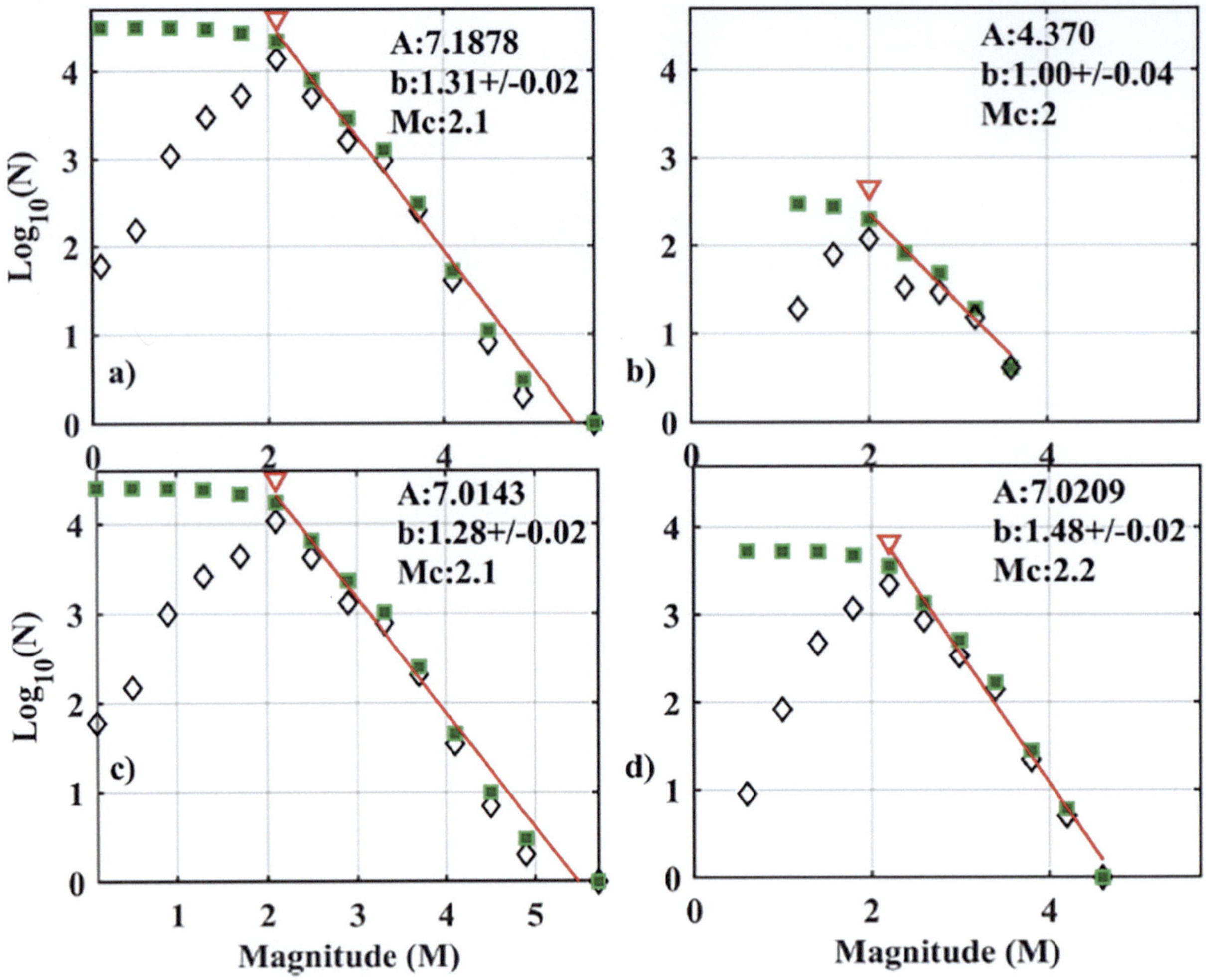

Fig. 8.5 Magnitude completeness (Mc) of catalog and b-value for the periods **a** January 2002–December 2018, **b** January 2002–December 2009, **c** January 2010–December 2016, and **d** January 2017–December 2018 computed using bin size 0.4. Selecting the higher bin value resolves the problems associated with magnitude uncertainties which are almost of the order 0.1–0.3. The diamond and square symbols denote the non-cumulative and cumulative number of earthquakes, respectively

8.3 Concluding Remarks

It may be concluded that a detailed analyses of the characteristics of injection induced earthquake sequences in Oklahoma, USA showed that:

1. There are similarities as well as differences in the characteristics of the injection induced seismicity (IIS) observed at Oklahoma, USA with global observations of reservoir triggered seismicity (RTS) and tectonic earthquakes.

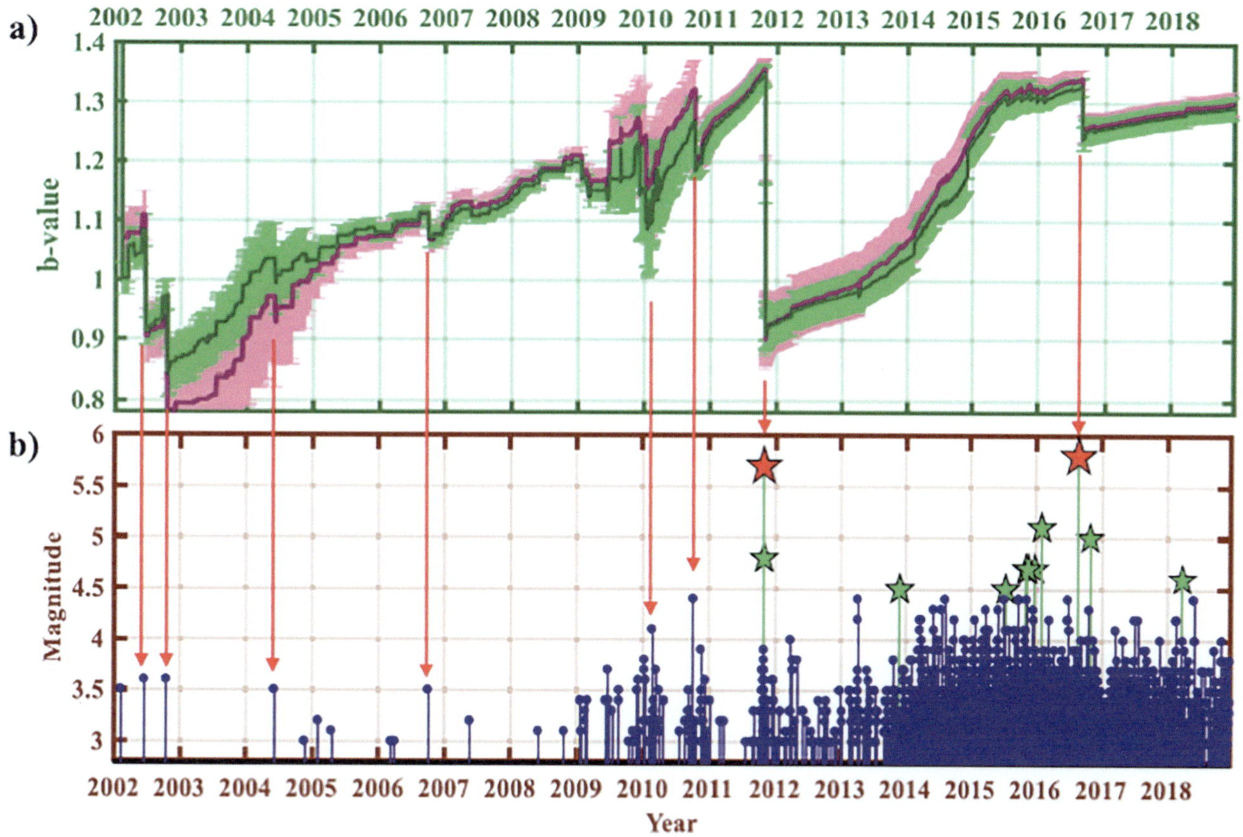

Fig. 8.6 Dark green and pink line plots in the top panel represent temporal variation of b-values computed using the expanding window of 6-h length from the seismicity data of Oklahoma for the period 2002–2018, respectively, with constant Mc (= 2.2) and threshold least square fit error minimization techniques. The standard error bars of green and pink lines are, respectively, shown in radium green and light pink colors. Bottom panel shows the seismicity. Green and red color stars in the bottom panel denote the earthquakes with M $\geq$ 4.5. Blue dots show earthquakes with M $\geq$ 3 and M < 4.5. Drop in b-value preceding the main shocks of different magnitudes is pointed with red arrows. Red color stars are Prague (2011, M 5.7) and Pawnee (2016, M 5.8) earthquakes selected to analyze earthquake sequence characteristics

The multiple injection wells operating from time to time in the region with varying volume of fluids injected make the IIS appear like an earthquake swarm.

2. A sharp fall in b values was observed before the M 5.7 Prague and M 5.8 Pawnee earthquakes like RTS earthquakes.
3. The difference 2 units of magnitude between the Pawnee earthquake and it's largest aftershock could be due to significant reduction in injection rates.
4. The Prague and Pawnee earthquakes occurred in the region with few mapped faults and in the vicinity of large paleo faults.

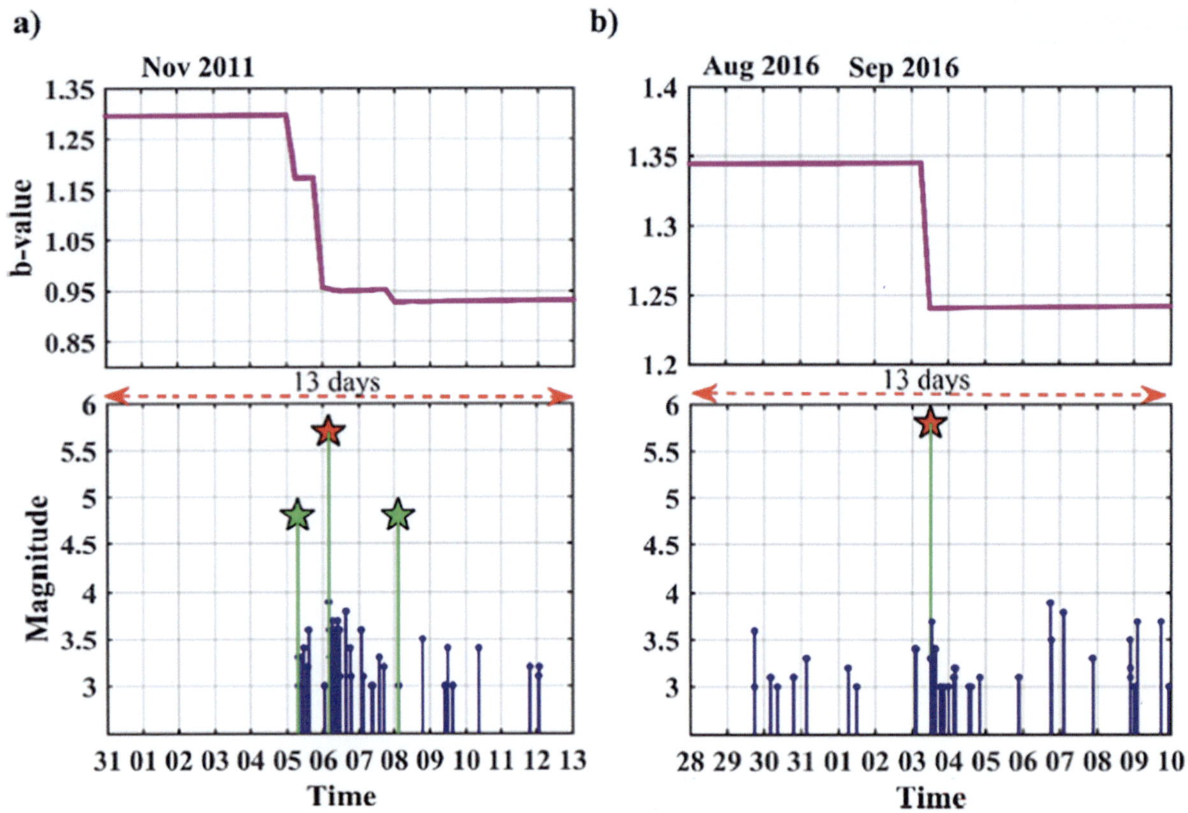

Fig. 8.7 Change in b-value prior to earthquakes: **a** 06 November 2011 (M 5.7), **b** 03 September 2016 (M 5.8). Green stars denote the foreshocks and aftershocks with $M \geq 4.5$. Blue dots denote the foreshocks and aftershocks with $M \geq 3$ and $M < 4.5$

Table 8.1 Characteristics of Prague (M 5.7) and Pawnee (M 5.8) earthquakes shown with red stars in Fig. 8.4

S. No.	Time of event	Epicenter		Depth (km)	M_0	$M_0 - M_1$	M_1/M_0	C	h	b-value		
	DD-MM-YY HH:MM	Latitude	Longitude							FS	AS	R
1	06-11-11 03:53	35.52° N	96.78° W	5.20	5.70	0.90	0.84	6.88	0.77	0.56 ± 0.07	0.96 ± 0.05	0.91 ± 0.01
2	03-09-16 12:02	36.43° N	96.93° W	5.60	5.80	2.00	0.66	5.50	1.00	0.66 ± 0.01	1.36 ± 0.02	0.91 ± 0.01

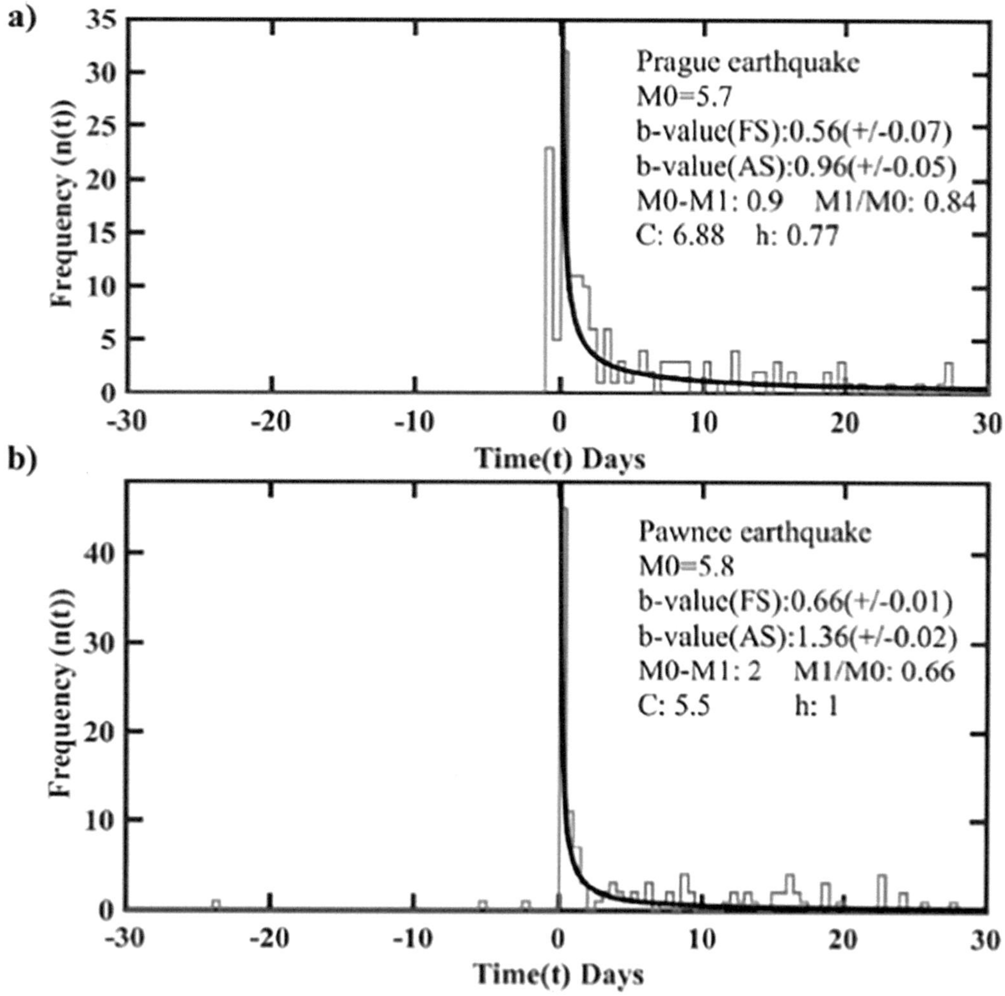

Fig. 8.8 Foreshock-aftershock pattern of **a** Prague and **b** Pawnee earthquake sequences plotted using unit time step 12 h within the circular region of radius of 15 km around the epicenter location

References

Bachmann CE, Wiemer S, Goertz-Allmann BP, Woessner J (2012) Influence of pore-pressure on the event-size distribution of induced earthquakes. Geophys Res Lett 39:L09302

Calais E, Han JY, DeMets C, Nocquet JM (2006) Deformation of the North American plate interior from a decade of continuous GPS measurements. J Geophys Res 111:B06402

Deichmann N, Giardini D (2009) Earthquakes induced by the stimulation of an enhanced geothermal system below Basel (Switzerland). Seismol Res Lett 80:784–798. https://doi.org/10.1785/gssrl.80.5.784

Ellsworth WL (2013) Injection-induced earthquakes. Science 341(6142):1225942

Evans MD (1966) Man made earthquakes in Denver. Geotimes 10:11–17

Gibbs JF, Healy JF, Raleigh CB, Coakley J (1973) Seismicity in the Rangely Colorado area: 1962–1970. Bull Seismol Soc Am 63:1557–1570

Gupta HK, Rastogi BK, Narain H (1972a) Common features of the reservoir-associated seismic activities. Bull Seismol Soc Am 62(2):481–492

Gupta HK, Rastogi BK, Narain H (1972b) Some discriminatory characteristics of earthquakes near the Kariba, Kremasta, and Koyna artificial lakes. Bull Seismol Soc Am 62(2):493–507

Gupta HK (1992) Reservoir Induced Earthquakes. Elsevier Scientific Publishing Company. Amsterdam. p 364

Hadsell FA (1968) History of earthquake activity in Colorado. Q Colo Sch Mines 63:57–72

Handin J, Raleigh CB (1972) Man-made earthquakes and earthquake control. In: Proceedings of the symposium on percolation through fissured rocks Deutsche Gesellschaft für Erd- und Grundbau, Stuttgart, FR Germany T 2D, pp 1–10

Healy JH, Rubey WW, Griggs DT, Raleigh CB (1968) The Denver earthquakes. Science 161:1301–1310

Horton S (2012) Disposal of hydro fracking waste fluid by injection into subsurface aquifers triggers earthquake swarm in central Arkansas with potential for damaging earthquake. Seismol Res Lett 83:250–260

Hough SE, Page M (2015) A century of induced earthquakes in Oklahoma? Bull Seismol Soc Am 105(6):2863–2870

Johann L, Shapiro SA, Dinske C (2018) The surge of earthquakes in Central Oklahoma has features of reservoir-induced seismicity. Sci Rep 8:11505

Keranen KM, Savage HM, Abers GA, Cochran ES (2013) Potentially induced earthquakes in Oklahoma, USA: links between wastewater injection and the 2011 Mw 5.7 earthquake sequence. Geology 41(6):699–702

Keranen KM, Weingarten M, Abers GA, Bekins BA, Ge S (2014) Sharp increase in central Oklahoma seismicity since 2008 induced by massive waste-water injection. Science 345(6195):448–451
Langenbruch C, Zoback MD (2016) How will induced seismicity in Oklahoma respond to decreased saltwater injection rates? Sci Adv 2(11):e1601542
Main IG, Meredith PG, Jones C (1989) A reinterpretation of the precursory seismic b-value anomaly from fracture mechanics. Geophys J Int 96(1):131–138
Mallika K, Gupta H, Shashidhar D, Rao NP, Yadav A, Rohilla S, Satyanarayana HVS, Srinagesh D (2013) Temporal variation of b value associated with M $\sim$ 4 earthquakes in the reservoir triggered seismic environment of the Koyna-Warna region, Western India. J Seismol 17(1):189–195
Mogi K (1962) Study of elastic shocks caused by the fracture of heterogeneous materials and its relation to earthquake phenomena. Bull Earthq Res Inst 40:125–173
Mogi K (1967) Earthquakes and fractures. Tectonophysics 5(1):35–55
Pakiser LC, Eaton JP, Healy JH, Raleigh CB (1969) Earthquake prediction and control. Science 166:1467–1474
Rajesh R, Gupta HK (2021) Characterization of injection-induced seismicity at north central Oklahoma, USA. Short communication. J Seismol 25:327–337
Raleigh CB (1972) Underground waste management and environmental implications. Am Assoc Pet Geol Mem 18:273–279
Schoenball M, Ellsworth WL (2017) A systematic assessment of the spatiotemporal evolution of fault activation through induced seismicity in Oklahoma and southern Kansas. J Geophys Res 122(12)
Simon RB (1969) Seismicity of Colorado: consistency of recent earthquakes with those of historical record. Science 165:897–899
Utsu T (1961) A statistical study on the occurrence of aftershocks. Geophys Mag 30(40):523–605
Van der Baan M, Calixto FJ (2017) Human induced seismicity and large scale hydrocarbon production in the USA and Canada. Geochem Geophys Geosyst 18. https://doi.org/10.1002/2017GC006915
Van der Elst NJ, Page MT, Weiser DA, Goebel THW, Hosseini SM (2016) Induced earthquake magnitudes are as large as (statistically) expected. J Geophys Res Solid Earth 121:4575–4590. https://doi.org/10.1002/2016JB012818
Walter JI, Ogwari P, Thiel A, Ferrer F, Woelfel I, Chang JC, Darold AP, Holland AA (2019) The Oklahoma geological survey statewide seismic network. Seismol Res Lett 91(2A):611–621
Zoback MD (2012) Managing the seismic risk posed by wastewater disposal. Earth Mag 57:38–43

9 Way Forward

We wish to start this chapter by quoting Zoback et al. (2007): "*The objective of fault-zone drilling projects is to directly study the physical and chemical processes that control deformation and earthquake generation within active fault zones. An enormous amount of field, laboratory, and theoretical work has been directed towards the mechanical and hydrological behavior of faults over the past several decades. Nonetheless, it is currently impossible to differentiate between—or even adequately constrain—the numerous conceptual models of active faults proposed over the years. For this reason, the Earth science community is left in the untenable position of having no generally accepted paradigm for the mechanical behavior of faults at depth. One of the primary causes for this dilemma is the difficulty of either directly observing or inferring physical properties and deformation mechanism along faults at depth, as well as the need to observe directly key parameters such as the state of stress acting on faults at depth, pore fluid pressure (and its possible variation in space and time), and processes associated with earthquake nucleation and rupture. Today, we know very little about the composition of active faults at depth, their constitutive properties, the state of in situ stress or pore pressure within fault zones, the origin of fault-zone pore fluids, or the nature and significance of time-dependent fault-zone processes*".

Hopefully, the study of triggered earthquakes at Koyna, India and elsewhere shall ultimately provide information on the following important issues:

1. Fluid pressure and permeability within and adjacent to the fault zone.
2. Composition and origin of fault-zone fluid and gasses.
3. Variation of the stress orientation and magnitude across the fault
4. Nucleation of earthquakes.
5. Propagation of earthquake rupture.
6. The role of water reservoirs in triggering the earthquakes.

It would be a great contribution to understand the genesis of earthquakes through the study of RTS, globally.

9.1 KTB Borehole

One of the earliest significant approaches to unravel the physical conditions of rocks at depth was made by the German Continental Deep Drilling Program, known as the KTB borehole. The project was carried out during 1987 through 1995 near Windischeschenbach, Bavaria in Germany (Emmermann 1997). The project included drilling of a pilot borehole of 4 km depth followed by the main borehole of 9.1 km depth. The pilot borehole was spudded 22 September 1987 and completed on 4 April 1989. The main borehole was spudded on 6 October 1990, and it reached a depth of 9.1 km on 21 October 1999. Funded by the Federal Ministry of Research of Germany, the project cost was some 270 Million Euros. A deep seismic observatory was installed which functioned from 1996 to 2001. As summarized by Emmermann and Lauterjung (1997), the major objectives of the program were: (1) the nature of geophysical structures and phenomena, (2) the crustal stress field and the brittle-ductile transition, (3) the thermal structure of the crust, (4) crustal fluids and transport processes, and (5) structure and evolution of the central European Variscan basement. The derrick used to drill the main borehole is one of the biggest, globally, and has been retained for demonstration purpose. It has become a tourist attraction. The boreholes are maintained for testing and experimentation.

The KTB project faced high friction and vertical instability. The drill head was designed to work in the anticipated temperature range of 250–300 °C at depths of 10–14 km.

H. K. Gupta, *Artificial Water Reservoir Triggered Seismicity*, https://doi.org/10.1007/978-3-032-05873-7_9

At the chosen location, KTB was expected to encounter *Erbendorfkörper*, which is supposed to be the boundary of a former continental plate. However, *Erbendorfkörper* was not reached. The temperatures encountered in KTB were much higher than anticipated. This issue of finding higher temperatures was much debated and gave rise to new approaches to comprehend the high temperatures encountered in KTB. Another surprise was that it was anticipated that the high temperatures and pressures would give rise to metamorphism. But that was not the case and instead large amounts of gases and fluids were encountered at those depths. This resulted in changing the drilling strategies for later deep bore holes. Using the KTB data the seismic reflections from depth were better interpreted; Dipole-Dipole Experiment measured the electrical conductivity around the borehole and many more.

One of the major outcomes of the KTB program was the founding of the German Research Centre for Geosciences (GFZ) in 1992. And in 1996, based at GFZ, the International Continental Scientific Drilling Programme (ICDP) was set up. Globally, ICDP has made an impact in learning more about the Earth's interior through drilling and observations in the drilled wells. As a matter of fact, ICDP contributed very significantly to the establishment of the 3 km deep Pilot Borehole at Koyna, India (Gupta 2011, 2018; Gupta et al. 2014, 2016).

While celebrating 30 years of reaching the final depth of the borehole on 12 October 1994 in 2024, the major contributions of KTB project were summarized as (quoted from ICDP Website):

- *on geophysical structures and phenomena*:
  seismic reflectors, electrical, magnetic and gravimetric anomalies; calibration of deep sounding methods
- *on the Earth's stress field and the transition from brittleness to ductility*:
  orientation and magnitude of mechanical stress as a function of depth; generation and triggering of earthquakes
- *on the thermal structure of the Earth's crust*:
  temperature distribution, heat flow, heat production; geothermal energy
- *on fluid and transport processes*:
  fluid sources, fluid pathways and budget; occurrence and distribution of mineral deposits
- *on the structure and development of the Paleozoic crust of Central Europe*:
  characteristics, deformation mechanism and dynamics of a reactivated crust.

9.2 San Andreas Fault Observatory at Depth (SAFOD)

The San Andreas Fault Observatory at Depth (SAFOD) project started in 2002 with an aim to collect necessary data from the San Andreas Fault enabling understanding the genesis of earthquake occurrence and possibly forecasting future earthquakes (Ellsworth et al. 2005). The project started in 2002. The pilot borehole was drilled to a depth of 2.2 km and the main borehole to a depth of 3.2 km. The drilling operations were completed in 2007. The major outcome of this project is presented by Zoback et al. (2007). SAFOD is located near Parkfield town in California, USA. One of the major achievements of SAFOD was placing of geophone and GPS clocks in the boreholes passing through the fault zones. These placements of a variety of geological and geophysical sensors and measurements on the samples collected from depth, provided invaluable data on the mechanical and geochemical properties of the rocks. SAFOD was funded under the Earthscope Program supported by the National Science Foundation of USA and the U S Geological Survey. The two boreholes were successfully drilled through the San Andreas Fault. There were some issues about the stability of the boreholes. The project has been successful having provided information about the state of stress, heat flow, gas contents within the fault zone. The findings of SAFOD can be listed as follows (after SAFOD: Wikipedia):

Wellbore Stability and Mud Penetration

1. As indicated by the caliper and hole volume data, the SAFOD boreholes were drilled successfully with little failure.
2. An increase in the size of the borehole with time was deciphered by the six-arm caliper. This may have happened due to mud-penetration.
3. The resistivity logs and finite element modeling further confirmed the process of slow mud penetration.

State of Stress and Fault Behavior

1. Significant local variations in the maximum horizontal compressive stress (SHmax) were inferred from the borehole breakouts and drilling induced fractures.
2. A transition from strike-slip to reverse faulting with depth was inferred from the increase in the angle between SHmax and San Andreas Fault.

3. High horizontal differential stress was inferred from the simultaneous occurrences of tensile fractures and borehole breakouts.

Heat Flow

1. Compared to the Parkfield area, where the heat flow at the basement is 74 m W m^{-2}, a higher heat flow of 91 m W m^{-2} was found in SAFOD.
2. Across the fault, no changes in heat flow were found, implying that the fluid flow has not altered the thermal regime significantly.

Analysis of Mud Gas

1. Gas logging in real time was carried out successfully in the sedimentary and granitic sections of the borehole.
2. Sedimentary section had higher mud gas concentration for Methane, Radon, and Carbon Dioxide.
3. Basement section had higher Helium concentration compared to the sedimentary section.
4. Two of the three shear zones had increased mud gas concentration.
5. Zones of interest for sampling were identified through mud gas analysis.

Other Important Findings

1. Internal structure and lithology of the San Andreas Fault was investigated. A 2.6 m thick clay gauge was discovered.
2. Friction rate parameters were found to vary with sliding velocity and normal stress through the studies of frictional properties.
3. Earthquake nucleation, rupture propagation and arrest to be monitored by appropriate seismic instrumentation deployed in the fault zone.

In an overview study of SAFOD, Zoback et al. (2011) have observed the San Andreas Fault to be relatively broad, about 200 m, which contains several discrete 2–3 m wide zones that exhibit low P- and S-wave velocities and low resistivity. They conclude that the SAFOD experiment supported strong crust weak fault model for the San Andreas Fault.

Research, using the results from the SAFOD experiment continues. In one of the important publications, Kaneko et al. (2017) have modeled the nucleation process of an M 2 earthquake. They observe that the rupture of most of the earthquakes starts abruptly, and there is no evidence of the nucleation process. However, several well studied earthquakes are found to have slow slip preceding the occurrence of the earthquake. However, observation of the slow slip remains a challenge. Kaneko et al. (2017) have made use of data acquired at SAFOD to develop computer models of the nucleation process of earthquakes on the San Andreas Fault. They predicted what the nucleation process will look like for Magnitude-2 earthquakes at 2.8 km depths at the San Andreas Fault. It is believed that the nucleation phase that has not been detected, may be observed in a few hours before the occurrence of an M 2 earthquake with instruments located close to the hypocenter. This would greatly help in understanding the physics of earthquake occurrence.

9.3 Reservoir Triggered Seismicity

A global review of RTS indicates that at least at five sites RTS events of $M \geq 6$ have occurred. Depth of the water column in the reservoir, identified by Stuart-Alexander and Mark (1976) continues to be the most important correlate for the occurrence of RTS. Earlier Gupta et al. (1972a, b) had identified certain common characteristics of RTS sequences that discriminated them from normal earthquake sequences. They inferred that under suitable geological conditions, filling of an artificial water reservoir influences the mechanical properties of the media, increases heterogeneity, and thereby earthquake sequences belonging to Type I of Mogi's (1963) model change to Type II. These characteristics are helpful in assessing whether an earthquake sequence belongs to RTS or not and have been frequently used. Latest use of this approach has been made by Pavlou (2019) to establish that the 1995 M 6.6 in the vicinity of Polyphyto Dam, Greece, was an RTS event. The same approach, applied to the 2008 Sichuan earthquake, which has been debated for a long time to have been triggered by the nearby Zipingpu reservoir, has shown that it was not a case of RTS (Gupta and Rekapalli 2025).

A review has been made of the work carried out to address the mechanism of induced seismicity. It is found that the changes brought about by loading/unloading of a reservoir in the stress field are very small, and the region must be stressed close to critical for these small changes to trigger earthquakes. Most of the energy released due to triggering is accumulated through tectonic processes.

Globally, the correspondence in the water-level variations and its influence on RTS is well established. However, due to the lack of knowledge of the physical properties of the rocks and fluids in the fault zone, the mechanism of RTS is not yet comprehended. As a matter of fact, globally, there is a lack of near-field studies. Koyna, India, provides an excellent site for such investigations. The RTS at Koyna is limited to an area of 20 × 30 km, the focal depths are mostly within 8 km, there is no other source of earthquakes within 100 km of Koyna Dam, the shallow crust is cool, with estimated temperatures not exceeding 130 °C at a depth of 6 km, the area being accessible for all kind of observations, and above all the RTS is still continuing. With the help of ICDP, workshops were organized for assessing the suitability of Koyna for putting borehole laboratory (Gupta 2018). A Pilot Borehole has been completed in 2017, and necessary measurements are being conducted for finalizing the specifications of the borehole laboratory at a depth of ~ 7 km. It is hoped that the outcome of the work at Koyna would not only elucidate RTS mechanism but would also provide crucial inputs to address several unanswered questions related to the genesis of earthquakes.

9.4 Work Carried Out at Koyna, India

The scientific work that started soon after the occurrence of the 10 December 1967 M 6.3 Koyna earthquake has provided significant information about the structural details of the region, the causative fault, Donachiwada, is well mapped. The 20 km × 30 km region was subjected to detailed airborne gravity and magnetic surveys, LiDAR mapping of the bare earth in the region, and magnetotelluric surveys to delineate the structural details. Nine boreholes going to a depth of ~ 1.5 km, penetrating the ~ 900 m basalt cover and going a few hundred meters in the granitic basement have provided immense geological-geophysical information about the region. The boreholes were geophysically and geologically logged and the cores have been subjected to various measurements. Setting up of a network of three component borehole seismometers below the basalt cover in the 8 boreholes has immensely improved the location capability of earthquakes occurring in the region. Based on the detailed investigation, the location of the 3 km deep Pilot Borehole in the Koyna region in the immediate vicinity of Donachiwada Fault, which had hosted Koyna earthquake of M 6.3 on 10 December 1967 and several M ~ 5 earthquakes was chosen (Gupta et al. 2017). The 3 km borehole was spudded in December 2016 and completed in June 2017. It was a totally successful drilling. Participants from all over the world and India participated in the post operation ICDP Workshop held in October 2017. They were totally satisfied with the drilling done, experiments and measurements carried out while drilling, collection of the cores and setting up of excellent facilities at Borehole Geophysics Research Laboratory (BGRL), Karad, close to the Koyna Dam site. In the years to follow, very detailed investigations have been carried out on the core samples in India and abroad. Several of these investigations are included in Chaps. 6 and 7 of the present book.

While analyzing core samples and continuing studies of earthquakes occurring in Koyna, COVID-19 paralyzed the operations for a couple of years. It is hoped that the necessary progress shall shortly be made to achieve the goals of the Koyna Deep Drilling Project.

9.5 Study of RTS Leads to Earthquake Forecast

RTS studies can also provide ways and means to forecast earthquakes. Gupta (2001) observed that short term earthquake forecast may be feasible in Koyna, India. The observation was based on the fact that earthquakes were occurring in a small area of 20 km × 30 km (Fig. 9.1), they were shallow, the area was covered by an excellent network of seismic stations (Fig. 9.2), there was no other source of earthquakes within 100 km of Koyna Dam, there was a direct correspondence between the water level in the reservoir and occurrence of earthquakes (Fig. 9.3). Moreover, a quasi-dynamic nucleation of smaller events preceeded before the occurrence of M ~ 4 earthquakes. For example, Fig. 9.4 exhibits the depth of the foreshocks deepening from 1 to 8.6 km in the preceding 60 h of the 26 April 1996 M 4.4 earthquake in the Koyna region. Recognition of precursory nucleation before M ~ 4 earthquakes led to successful forecasting of M ~ 4 earthquakes in the Koyna region (Gupta et al. 2006, 2007; Gupta 2011).

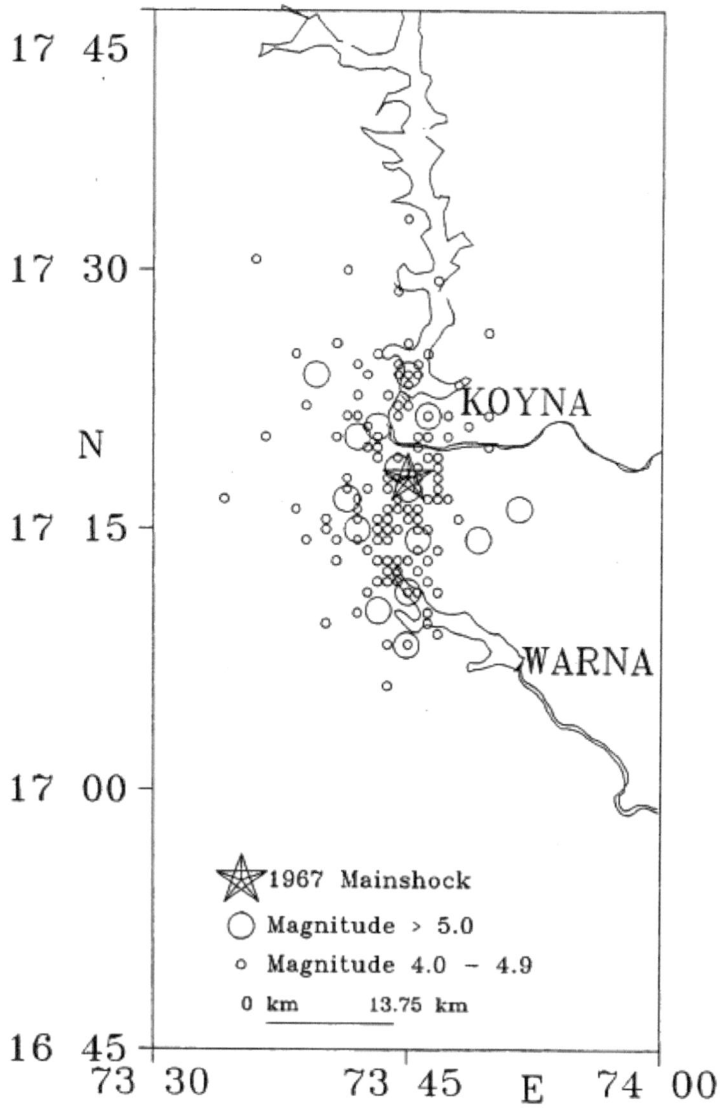

Fig. 9.1 Epicenters of earthquakes of magnitude 4 and larger for the period 1963 through 1980 in Koyna-Warna region near West Coast of India

In a recent paper Orlecka-Sikora et al. (2025) have addressed the dynamics of earthquake preparation process in reservoir triggered seismicity zones. They underline the complex interplay between anthropogenic activities and the natural geological process. Understanding of this complex interplay could be helpful in mitigating RTS generated hazard. They have experimented with their approach the RTS being observed at Song Tranh 2 Reservoir in Vietnam and Assu Reservoir in Brazil. They have developed a five-stage damage model that includes (Orlecka-Sikora et al. 2024). They observe that large earthquakes are often preceded deformation. The entire process consists of the following 5 stages:

1. Primary Stage: Compaction stage where initial fractures form, but they are isolated.
2. Secondary Stage: Quasi-static growth of the fractures.
3. Nucleation Stage: Initiation of macrocracks, coalescing of fractures, increase in stress concentration, and possible regularization of fracture spacing.
4. Hardening Stage: Strengthening of rocks, dislocation movement becomes harder.
5. Failure Stage: Slipping of fault, with closing of the microcracks.

The most important issue is to be able to identify these stages in earthquake preparation zones.

We quote Orlecka-Sikora et al. (2024): "*Observations and experiments indicate that large earthquakes are often preceded by significant deformation in the rock mass. The key challenge is identifying reliable geophysical signals that can help predict when these changes are occurring. By combining insights from studies on rock fracturing with real-world data from human-induced seismicity, we identified*

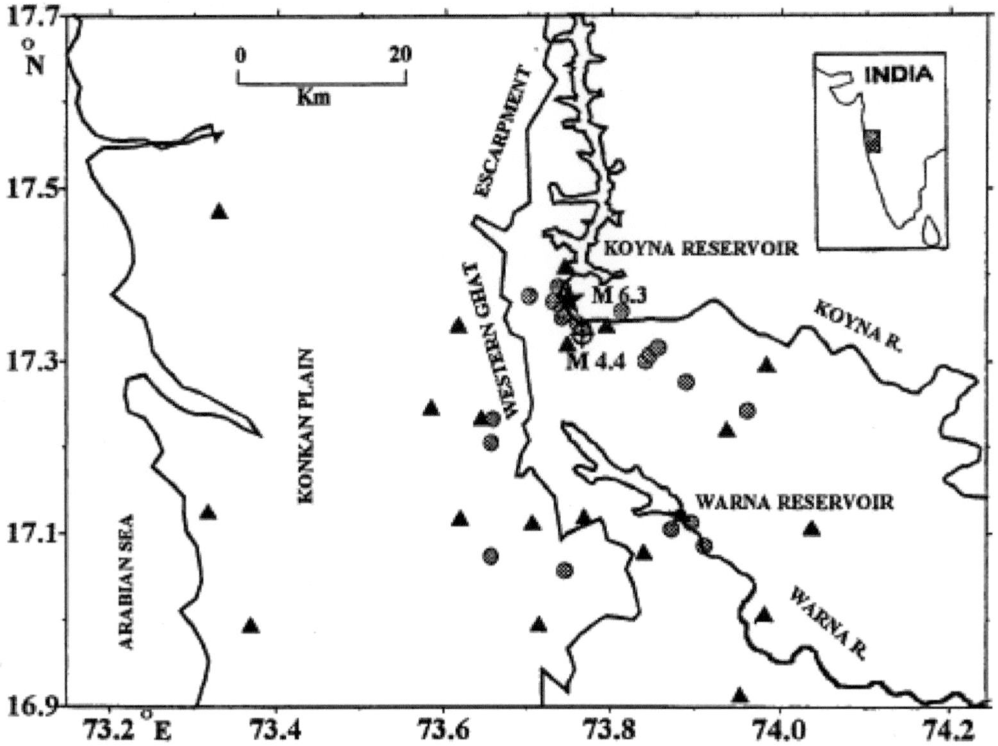

Fig. 9.2 The location of Koyna-Warna reservoirs in west India. Triangles indicate the sites where seismic stations are operated. Circles show the locations of borewells. The largest triggered earthquake of M 6.3 on December 10, 1967 is shown by a star. A circle and a plus sign show the M 4.4 earthquake of April 25, 1997, referred in Fig. 9.4

critical parameters that track the progression of earthquake-related damage. These parameters are tied to a five-stage model of rock damage, which helps assess the current state of the system and the likelihood of a major earthquake. Using laboratory data, we examined how these parameters behave under stress and damage conditions, revealing how cracks grow and cluster over time. The model was applied to ten major earthquakes caused by activities such as dam construction, mining, and geothermal energy production. Our findings show that gradual damage consistently precedes these events, with larger earthquakes taking longer to reach the final stage of damage. We also identified two distinct types of preparatory earthquake behavior: a faster one associated primarily with mining activities and a slower one linked to fluid-related cases. This understanding could help mitigate earthquake risks in energy projects through targeted interventions".

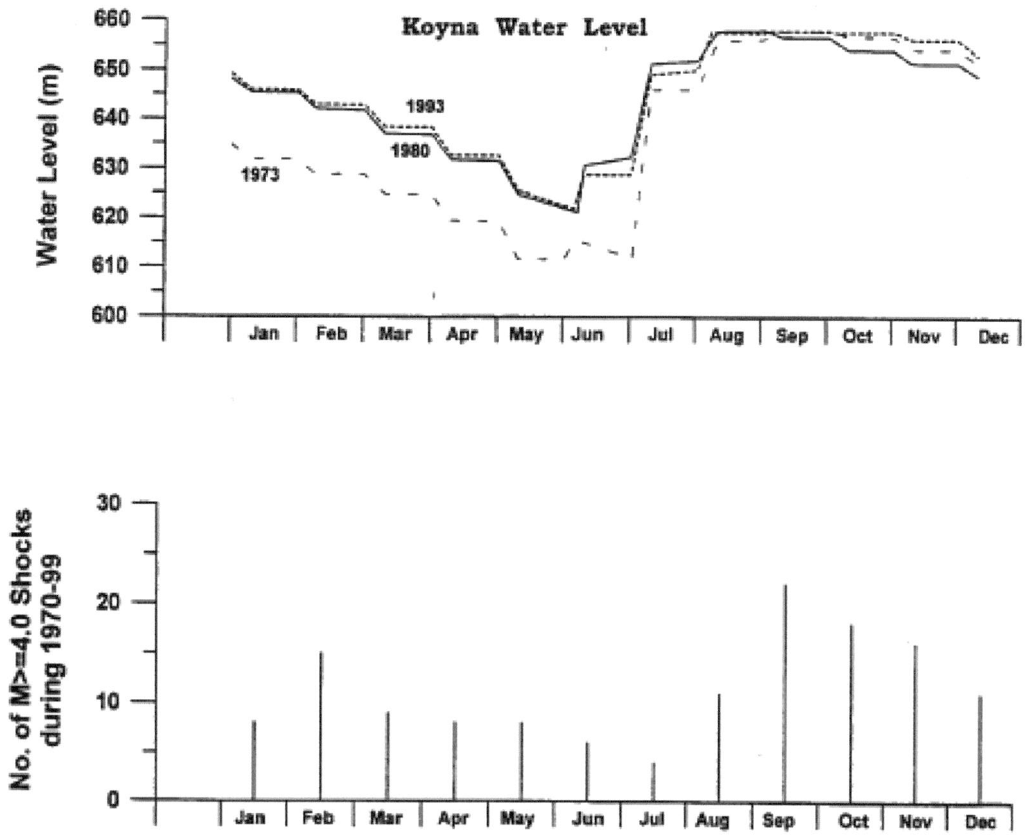

Fig. 9.3 Water levels in Koyna reservoir for the years 1973, 1980 and 1993 and the monthly number of earthquakes of M 4 and larger for the period 1970 through 1999

9.6 Importance of Deep Borehole Investigations at RTS Sites

As stated by Zoback et al. (2007), "*The objective of fault-zone drilling projects is to directly study the physical and chemical processes that control deformation and earthquake generation within active fault zones.*" However, they also take cognizance that despite all the efforts made in this direction, we know precious little about what exactly happens in the fault zones while they are getting ready to nucleate an earthquake. Making appropriate observations to learn the fault behavior in complex earthquake belts like the San Andreas Fault or the Himalayan Earthquake Belt is very difficult. These earthquake zones extend to thousands of kilometers, earthquakes occur at depths of a few kilometers to tens of kilometers, the earthquake zones are not easily accessible to make necessary observations and measurements due to their geographical locations. It is very difficult to define what *causes* earthquakes and what is the *effect* of any observable physical intervention in *causing* of an earthquake. In this context, Koyna, India, was found to be an ideal site for undertaking studies in the fault zone of the M 6.3 earthquake which also hosted several M ~ 5 earthquakes.

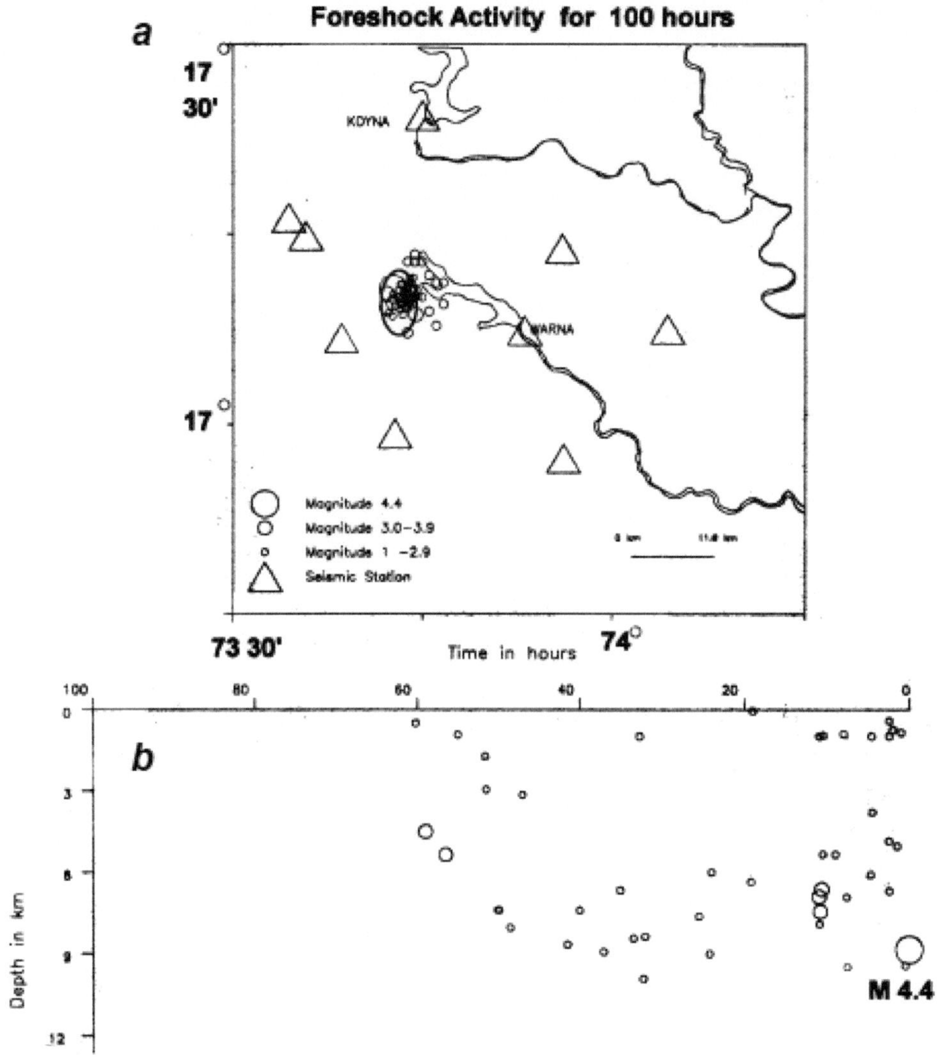

Fig. 9.4 **a** Epicentres of foreshocks during 6–26 April 1996 for the Koyna mainshock of M 4.4. The foreshocks within an area of 8 km × 6 km are marked by an ellipse. **b** Depth variation of foreshocks with time. The mainshock epicentre is shown by the bigger circle, $M = 3$–3.9 by medium-sized circles and $M < 3$ by small circle (after Rastogi and Mandal 1999)

A majority of RTS sites are in the stable continental regions. Normally, a dam would be built in a location which is easily accessible for construction and other developmental activities. It has been also observed that RTS events are mostly shallow. For example, the RTS at Hsinfengkiang, China, was observed between the depths of 1 and 10 km with highest concentration at 6 km depth (Fig. 9.5). Similarly at Koyna, India most events occurred between a depth of 2 and 10 km with highest concentration at around 5 km. Earthquakes close to Koyna Dam had a depth to 10 km. However, seismicity shoaled towards south of the Koyna Dam (Fig. 9.6). Similarly, RTS at the Lake Kariba on Zambia-Zimbabwe border was confined from 4 to 10 km as can be seen from Table 9.1 after Pavlin and Langston (1983).

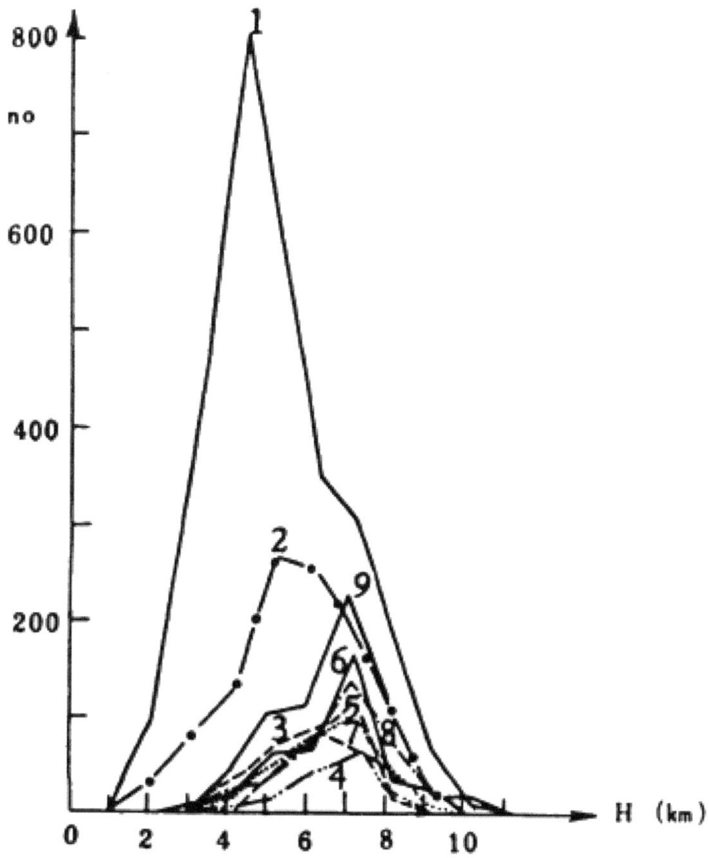

Fig. 9.5 Focal depth distribution for various periods for earthquakes of Ms > 1.0 in Hsinfengkiang region. 1: March 19, 1962–July 23, 1962; 2: July 24, 1962–November 27, 1962; 3: November 28, 1962–April 14, 1963; 4: April 15, 1963–August 8, 1963; 5: August 9, 1963–December 14, 1963; 6: December 15, 1963–April 19, 1964; 7: April 20, 1964–August 24, 1964; 8: August 25, 1964–December 29, 1964; 9: December 30, 1964–May 5, 1965

At Oroville Dam, California, USA, an RTS event of M 5.7 occurred on 1 August 1975. Figure 9.7 provides a depth section of Oroville earthquake sequence. Most of the events are located at depths of 3–10 km. Many more similar examples can be sighted.

The main objective of the discussion is that RTS sites have easy access, and the seismicity is mostly limited to shallow depths of 2–10 km. There is a relation between the fluctuation of the water level and the occurrence of RTS. When an artificial water reservoir site is located in a stable continental region, there are no other earthquake sources in the nearby region to complicate comprehending the observed RTS. Therefore, these sites provide an excellent opportunity to carry out near-source studies. The work carried out at Koyna, India has demonstrated that focused research and observations using multi-disciplinary studies are likely to provide an answer to the questions facing seismologists all over the world as what really happens before an earthquake is nucleated and the rupture propagates. With the recent geotechnical and analytical developments, and the success of deep borehole observations at SAFOD, USA, and Koyna, India.

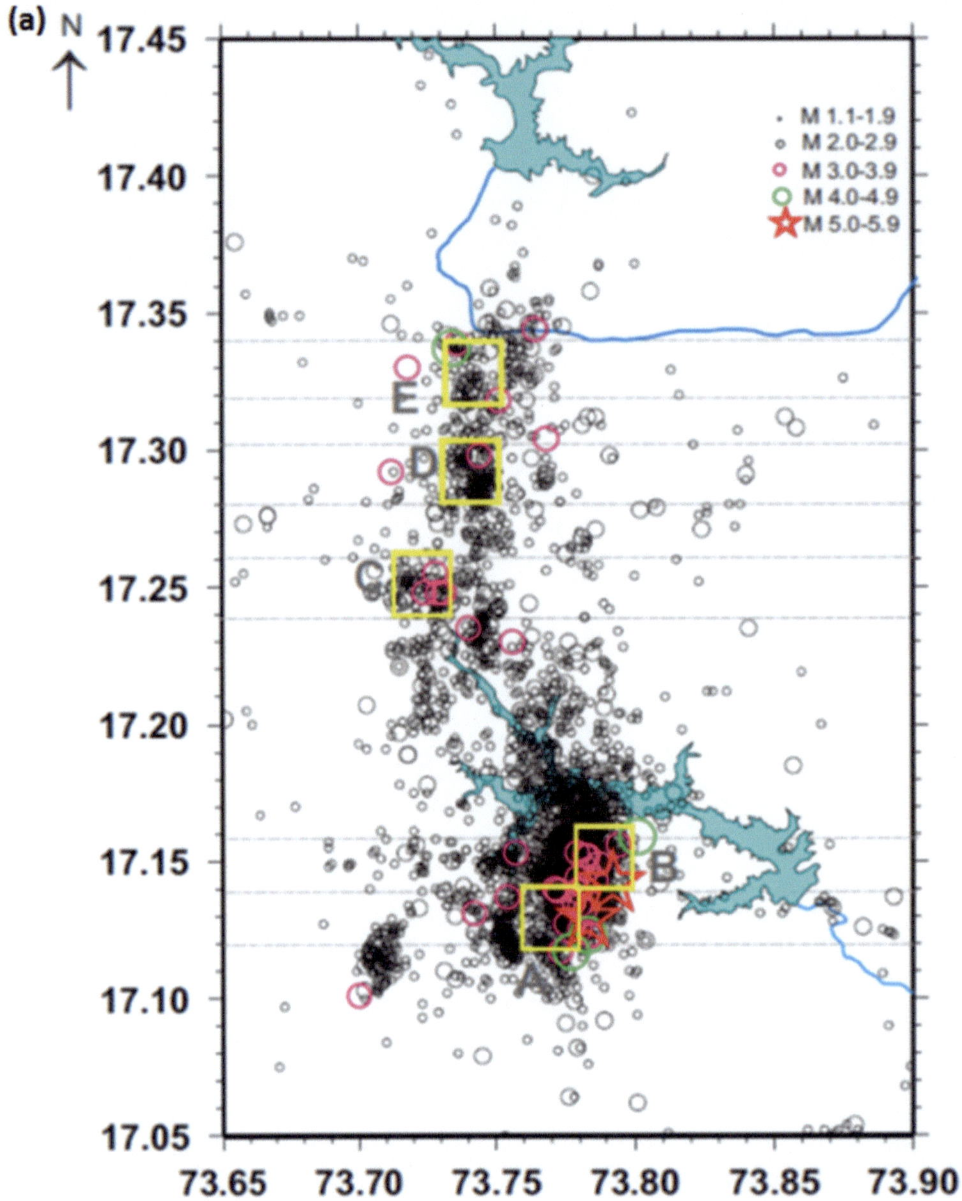

Fig. 9.6 **a** Seismicity of the Koyna-Warna region during 2009–2014. The five blocks A, B, C, D and E are each 2 × 2 km^2 in area. Dotted lines indicate a swath of 2 km area at each block. **b** Depth sections for 2 km swath for the 5 blocks. Within brackets are indicated the number of earthquakes that occurred in each of these blocks

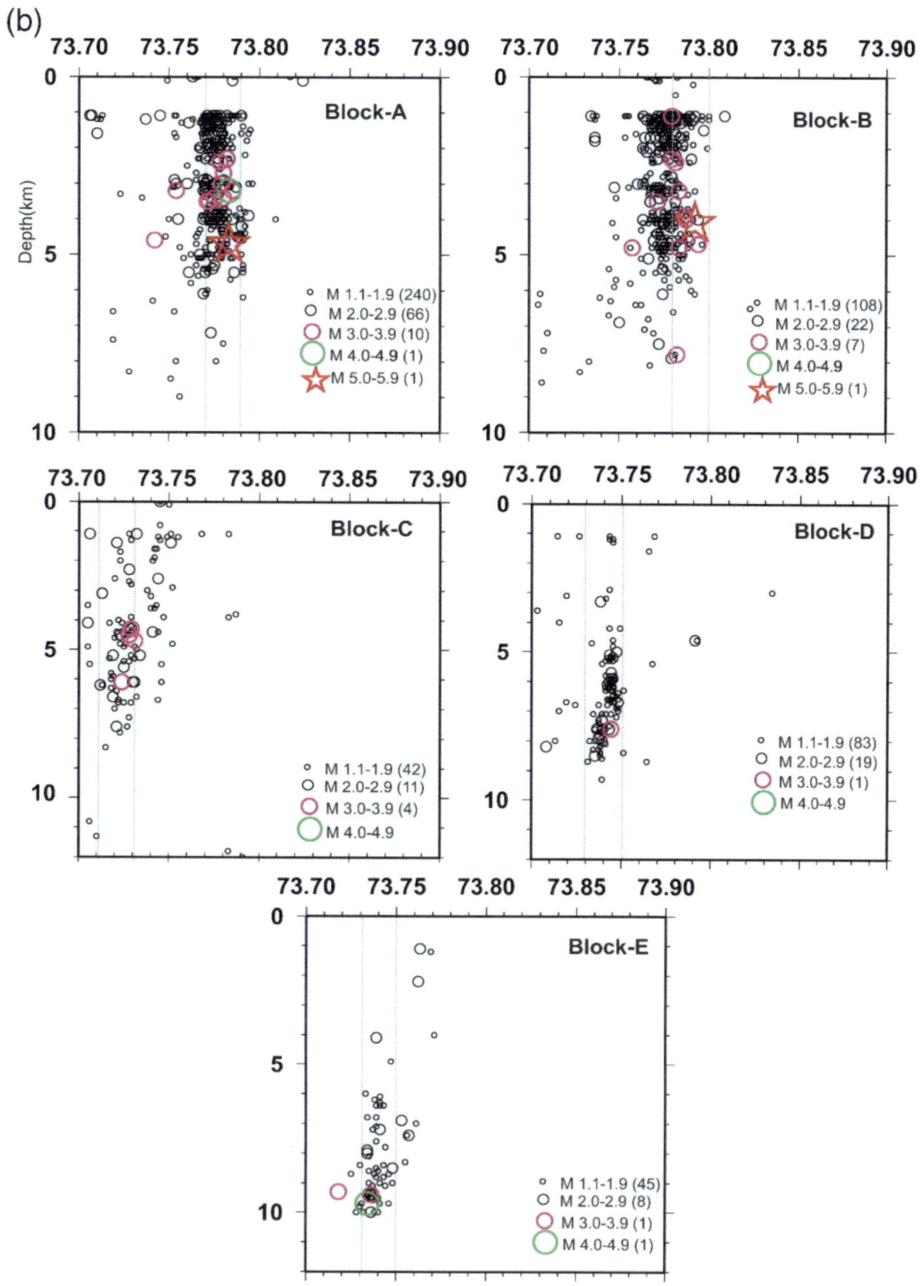

Fig. 9.6 (continued)

Table 9.1 Major earthquakes at Kariba (after Pavlin and Langston 1983)

Earthquake time (GMT)	Magnitude (m^)	Location	Depth[a]
0640 Sept. 23, 1963	5.6	16.69° S, 28.43° E	6.0
0810 Sept. 23, 1963	4.9	16.59° S, 28.70° E[b]	8.0
0901 Sept. 23, 1963	6.0	16.58° S, 28.46° E[b]	8.5
1502 Sept. 23, 1963	5.4	16.67° S, 28.40° E[b]	10.0
2223 Sept. 23, 1963	5.3	16.70° S, 28.42° E	5.0
0913 Sept. 24, 1963	5.3	16.64° S, 28.54° E[b]	4.0
0703 Sept. 25, 1963	5.8	16.71° S, 28.44° E	7.0
0959 Nov. 8, 1963	5.4	16.63° S, 28.56° E[b]	4.5
0251 Apr. 20, 1967	4.9	16.77° S, 28.37° E	4.5
0318 Dec. 12, 1972	5.2	16.74° S, 28.08° E	5.0
0118 Dec. 18, 1972	5.6	16.71° S, 28.07° E	5.0
0605 Jan. 13, 1973	5.3	16.78° S, 28.04° E	5.0
0936 Aug. 1, 1974	5.3	16.63° S, 27.89° E	5.5

[a] Depths derived from analyzing surface waves and short-period body waves
[b] Data from Fairhead and Girdler (1969)

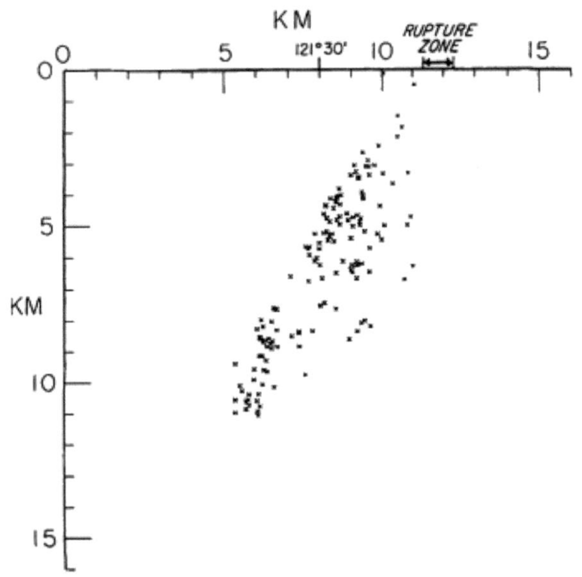

Fig. 9.7 An east–west depth section of the Oroville August 1975 earthquake sequence and the surface rupture zone (after Clark et al. 1976)

References

Clark MM, Sharp RV, Castle RO, Harsh PW (1976) Surface faulting near Lake Oroville, California in August, 1975. Bull Seismol Soc Am 66(4):1101–1110

Ellsworth W, Hickmann S, Zoback MD, Davis E, Gee L, Huggins R, Krug R, Lippus C, Malin P, Neuhauser D, Paulsson B, Shalev E, Vajapeyam B, Weiland C, Zumberge M (2005) Observing the San Andreas Fault at depth. Eos Trans Am Geophys Union 86(53). Fall meeting supplement, abstract T24B-04

Emmermann R (1997) The KRB pilot hole: tectonic setting, technical data and first results. Springer Nature, pp 527–553

Emmermann R, Lauterjung J (1997) The German continental deep drilling program KTB: overview and major results. J Geophys Res Solid Earth. https://doi.org/10.1029/96jB03945

Fairhead JD, Girlder RW (1969) Research note: the Kariba earthquake. Nature 221:1018–1020

Gupta HK (2001) Short-term earthquake forecasting maybe feasible at Koyna, India. Tectonophysics 338(3–4):353–357

Gupta HK (2011) Planning a deep drilling project in the Koyna region of India, workshop on deep scientific drilling to study reservoir triggered earthquakes, Koyna, India, 21–25 March 2011. Eos 92:283

Gupta HK (2018) Review: reservoir triggered seismicity (RTS) at Koyna, India over the past 50 years. Bull Seismol Soc Am 108(5B):2907–2918

Gupta HK, Rekapalli R (2026) The 2008 Mw 7.9 Wenchuan, China earthquake: not a case of reservoir triggered seismicity. Gondwana Research 151:184–188

Gupta HK, Rastogi BK, Narain H (1972a) Common features of the reservoir associated seismic activities. Bull Seism Soc Am 62:481–492

Gupta HK, Rastogi BK, Narain H (1972b) Some discriminatory characteristics of earthquakes near the Kariba, Kremasta and Koyna artificial lakes. Bull Seism Soc Am 62:493–507

Gupta HK et al (2006) Prediction of an M~4 earthquake in the Koyna region comes true! J Geol Soc India 68:149–150

Gupta HK et al (2007) Earthquake forecast appears feasible at Koyna, India. Curr Sci 93(6):843–848

Gupta HK, Purnachandra Rao N, Roy S, Kusumita Arora VM, Tiwari BPK, Patro HVS, Satyanarayana D, Shashidhar KM, Akkiraju VV, Goswami D, Vyas D, Ravi G, Srinivas KNSSS, Srihari M, Mishra S, Dubey CP, Raju CV, Ujjal Borah K, Reddy C, Babu N, Rohilla S, Dhar U, Mrinal Sen YJ, Rao B (2014) Investigations related to scientific deep drilling to study reservoir-triggered earthquakes at Koyna, India. Int J Earth Sci 104:1511–1522

Gupta HK, Kusumita Arora N, Rao P, Sukanta Roy VM, Tiwari PK, Patro HVSS, Shashidhar D, Mahato CR, Srinivas KNSSS, Srihari M, Satyavani N, Srinu Y, Gopinadh D, Raza H, Jana M, Akkiraju VV, Goswami D, Digant Vyas CP, Dubey DCV, Raju UB, Kashi Raju K, Reddy C, Narendra Babu BK, Bansal SN (2016) Investigations of continued reservoir triggered seismicity at Koyna, India. Geol Soc Lond Spec Publ 445:151–188

Gupta HK, Shashidhar D, Mahato CR, Satyanarayana HVS, Mallika K, Purnachandra Rao N, Maity BS, Navitha K (2017) Location of the pilot borehole for investigations of reservoir triggered seismicity at Koyna, India. Gondwana Res 42:133–139

Kaneko Y, Carpenter BM, Nielsen SB (2017) Nucleation process of magnitude 2 repeating earthquakes on the San Andreas Fault predicted by rate- and-state fault models with SAFOD drill core data. Geophys Res Lett. https://doi.org/10.1002/3016GL071569
Mogi K (1963) Some discussions on aftershocks, foreshocks and earthquakes swarms - the fracture of semi-infinite body caused by an inner stress origin and its relation to the earthquake phenomena. Bull Earthquake Res Inst 41:615–658
Orlecka-Sikora B, Ciechowska H, do Nascimento AF (2024) Insights into dynamics of earthquake preparation process in water reservoir-triggered seismic zones. J Geol Soc India 101(6):742–747
Orlecka-Sikora B, Ciechowska H, Zhang Z, Liang Z (2025) Seismic strain metrics for tracking fault damage evolution and rupture potential in anthropogenic seismicity. Rock Mech Rock Eng
Pavlin GB, Langston CA (1983) An integrated study of reservoir induced seismicity in the vicinity of Lake Bhatsa, Maharashtra, India. Phys Earth Planet Inter 44:73–81
Pavlou K (2019) Relationship between observer seismicity and water level fluctuations in polyphyto dam area (North Greece). J Geogr Environ Earth Sci Int 21(2):1–10
Rastogi BK, Mandal P (1999) Foreshocks and nucleation of small- to moderate-sized Koyna earthquakes (India). Bull Seismol Soc India 89(3):829–836
Stuart-Alexander DE, Mark RK (1976) Impoundment-induced seismicity associated with large reservoirs. U.S. Geo Surv Open File Rep 76–770
Zoback MD, Hickman S, Ellsworth W (2007) The role of fault zone drilling. In: Kanamore H, Schubert G (eds) Earthquake seismology-treatise on geophysics, vol 4. Elsevier, Amsterdam, pp 649–674
Zoback MD, Hickman S, Ellsworth WL (2011) Scientific drilling into the San Andreas Fault zone—an overview of SAFOD's first five years. Sci Drill. https://doi.org/10.2204/iodp.sd.11.02.2011